Big Wars and Small Wars

This book sets out to show how in the twentieth century the British army learnt lessons from one war in order to prepare for the next. The challenge was particularly great, as it never had the luxury of going directly from one major war to another, but always had to reckon with ongoing commitments to a range of 'small wars'. These included colonial campaigns between 1902 and 1914; counter-insurgency operations between 1945 and 1969; and peace support operations after 1991. The army believed that by preparing for major war it also enabled itself to prepare for lesser conflicts, and that is still its current doctrine. This volume explores the historical dimension to this debate, and offers analyses by the most prominent experts in the field, including Hew Strachan, Edward Spiers, David French, Paul Cornish, Daniel Marston, David Benest, Simon Ball and Colin McInnes.

This book will be of great interest to students of military history and strategic studies in general, and of particular interest to students of the British army and British military doctrine.

Hew Strachan is Professor of the History of War, University of Oxford, Director of the Oxford Leverhulme Programme on the Changing Character of War, and a Fellow of All Souls College. His publications include *European Armies and the Conduct of War* (1983), *The Politics of the British Army* (1997), and the first volume of a trilogy on the First World War, *To Arms* (2001).

ROUTLEDGE SERIES: MILITARY HISTORY AND POLICY
Series Editors: John Gooch and Brian Holden Reid

This series will publish studies on historical and contemporary aspects of land power, spanning the period from the eighteenth century to the present day, and will include national, international and comparative studies. From time to time, the series will publish edited collections of essays and 'classics'.

Big Wars and Small Wars

The British army and the lessons of war in the twentieth century

Edited by **Hew Strachan**

Routledge
Taylor & Francis Group

LONDON AND NEW YORK

First published 2006
by Routledge
2 Park Square, Milton Park, Abingdon, Oxon OX14 4RN

Simultaneously published in the USA and Canada
by Routledge
270 Madison Ave, New York, NY 10016

Routledge is an imprint of the Taylor & Francis Group, an informa business

Transferred to Digital Printing 2009

© 2006 Hew Strachan

Typeset in Times New Roman by Taylor & Francis Books

British Library Cataloguing in Publication Data
A catalogue record for this book is available from the British Library

Library of Congress Cataloging in Publication Data
A catalog record for this book has been requested

ISBN10: 0–415–36196–6 (hbk)
ISBN10: 0–415–54504–8 (pbk)

ISBN13: 978–0–415–36196–5 (hbk)
ISBN13: 978–0–415–54504–4 (pbk)

Contents

Contributors

Simon Ball is Reader in Modern History at the University of Glasgow. His most recent book, *The Guardsmen*, was published by HarperCollins in 2004. He is currently writing an analytical history of British defence policy since 1945 for Cambridge University Press'. Previous publications are: *Bomber Bases and British Strategy in the Middle East, 1945–1949* (1991), *The Bomber in British Strategy: Doctrine, Strategy and Britain's World Role, 1945–1960* (1995), *The Cold War: An International History, 1947–1991* (1997), and *The Great Powers and the Division of Europe, 1943–1949* (1997) (CD-ROM).

David Benest has served with the British Army for the past thirty-three years. He has accumulated nearly five years of operational deployments to Northern Ireland, principally to West Belfast and South Armagh. He spent a further six years in the Ministry of Defence in support of counter-terrorist operations. His last appointment was as Director Defence Studies (Army) and Director of the Strategic and Combat Studies Institute (SCSI). He is currently serving within the Defence Academy of the United Kingdom and is researching strategic leadership and counter-insurgency.

Paul Cornish holds the Peter Carrington Chair in International Security at Chatham House, where he is head of the International Security Programme. He has served in the British Army and the Foreign and Commonwealth Office, and has held appointments at Cambridge University and King's College London. His research and writing interests include European security and defence, the ethics of the use of armed force, and arms control. His thesis on *British Military Planning for the Defence of Germany, 1945–50* was published in 1996. Dr Cornish is the authorised biographer of General Sir Miles Dempsey.

David French is Professor of History at University College London, where he teaches and researches modern British military history. He is the author of *Raising Churchill's Army: The British Army and the War against Germany, 1919–1945* (2000) and *Military Identities: the British Army, the British People and the Regimental System, c.1870–2000* (2005). He is currently writing a history of British military policy from 1945 to c.1970.

Daniel Marston is Senior Lecturer in War Studies at the Royal Military Academy, Sandhurst. He completed both his B.A. and M.A. in History at McGill University, Montreal, Canada and his D.Phil. in the History of War at Balliol College, Oxford. Previous publications include *Phoenix from the Ashes: The Indian Army in the Burma Campaign* (2003) which won the Templer Medal.

Colin McInnes holds a personal chair in the Department of International Politics at the University of Wales, Aberystwyth. He was formerly a lecturer in War Studies at the Royal Military Academy Sandhurst, Visiting Research Fellow at the Centre for Defence Studies, King's College London and Special Adviser to the House of Commons Defence Committee. His books include *NATO's Changing Strategic Agenda* (1990), *Hot War, Cold War: The British Army's Way in Warfare, 1945–95* (1996), and *Spectator Sport War: The West and Contemporary Conflict* (2002).

Edward M. Spiers is the Professor of Strategic Studies at the University of Leeds. He has written extensively on military history, including *Haldane: An Army Reformer* (1980), *The Army and Society, 1815–1914* (1980), *Radical General: Sir George de Lacy Evans, 1787–1870* (1983), *The Late Victorian Army, 1868–1902* (1992) and *The Victorian Soldier in Africa* (2004). He also edited *Sudan: The Reconquest Reappraised* (1998).

Hew Strachan is Chichele Professor of the History of War and Fellow of All Souls College, at Oxford University, where he also directs the Leverhulme Programme on the Changing Character of War. His recent books include *The Politics of the British Army* (1997), *The First World War: To Arms* (2001, the first volume of a planned trilogy), *The First World War: A New Illustrated History* (2003), and, as editor, *The British Army, Manpower and Society into the 21st Century* (2000).

Preface and acknowledgements

In March 2003, the Strategic and Combat Studies Institute organised a conference in Oxford, at the behest of the Director General of Development and Doctrine, Major-General Jonathan Bailey. Its title was 'The Proof of the Pudding', and its aim was to consider the relationship between lessons learnt from immediate past wars and ideas developed in preparation for the next war. The focus was on the British army in the twentieth century. The army has never had the luxury of emerging from one major European war and then preparing itself for the next. In between it has always had to reckon with ongoing commitments to a range of 'small wars'. They included colonial campaigns, after 1902 and again after 1918, and latterly what we would now call counter-insurgency conflicts and peace support operations.

The army's current doctrine is that by preparing for major war it also enables itself to prepare for lesser conflicts. That it is highly competent in regard to the latter is regularly demonstrated; that it has also managed to retain a comparable competence in 'high intensity operations' seemed to be shown during the Gulf War of 1991. Whether this will hold true for future high intensity conflict remains to be seen. That will be the 'proof of the pudding'.

As the participants in the conference assembled they were conscious that the army itself was preparing for the invasion of Iraq in a second Gulf War. But that has proved a difficult pudding to digest. The spectacular early successes of the coalition apparently endorsed the doctrine. But it is hard to be certain. First, the nature of the fighting in the south of Iraq in April 2003, the operations in which the British army, as opposed to that of the United States, was principally involved, were not sustained at the same high tempo for as long as those around Baghdad and further north. Second, although President Bush soon declared the war to be over, fighting continued. The comparison with the war that begins this book is tempting. The British captured Pretoria in 1900: the guerrilla war against the Boers continued until 1902. How should these later stages be defined and assessed? The current British doctrine speaks of 'three-block' wars, in which high intensity operations, counter-insurgency and post-war settlement are all going on at the same time. If that is the case, learning the lessons from the second Gulf War is also still ongoing. Certainly, the Chief of the General Staff has not seen fit to publish those lessons, and they therefore cannot form part of this volume, which effectively ends with the close of the twentieth century.

This book is caught, as learning lessons must be, between whether its role is to discuss the past or to prescribe for the future. In choosing my tenses in writing the introduction, I have been conscious that at some points where I have used the perfect the imperfect or even the present could have been substituted. The reader needs to be aware of that. But, if in doubt, I have opted for the past tense, as the focus is on the historical dimension to the debate. Did fighting in the two world wars make the army more or less competent in its handling of 'small wars'? Did colonial campaigns and counter-insurgency aid the army's fitness for major war or undermine it? The essays which follow address these questions. Two have been added since the conference. Simon Ball considers the lessons from the Falklands War. Most problematic was the handling of the army's experience in Northern Ireland – which straddles a full third of the period under review but which remains an operation of considerable sensitivity and is therefore hard to write about. David Benest has filled the breach within the limits set by what can be in the public domain.

He – alongside General Bailey and Brigadier Mungo Melvin, then Director Land Warfare – was the progenitor of this book. As Colonel Defence Studies, David Benest was responsible for the organisation of the conference and its undoubted success.

Hew Strachan

Introduction

The essays in this book span the twentieth century. In its first half the British army began each of its major wars with defeat. In 1899 it was humiliated by the Boers, and not until 1902 did it finally overcome the Afrikaner republics; in 1914, it was driven back by the German advance through France and Belgium, and its conduct of every battle on the western front thereafter was dogged by controversy until the last 'hundred days' of 1918 itself; and between 1939 and 1942 the German army – as well as the Japanese – routed it whenever they met. In the second half of the century its wars were shorter, and it became more adept at achieving early success. By the year 2000 the army was deemed to be efficient, professional and effective. Its reputation would have amazed earlier commentators. For most of the previous hundred years the army's public image owed more to Colonel Blimp than to Bernard Montgomery or Bill Slim.

The more sophisticated of its critics did not just accuse it of being amateur, of preferring to muddle through – rather than to think through – its problems. They also accused it of being a prisoner of its own history. It had learnt lessons from its past wars but it had learnt them too well, and therefore failed to recognise what was new and changing. 'All the men who filled the highest commands in our Army in France were veterans of the Boer War', wrote David Lloyd George in the conclusion to his memoirs of the First World War.

> It is not too much to say that when the Great War broke out our Generals had the most important lessons of their art to learn. Before they had much to unlearn. Their brains were cluttered with useless lumber, packed in every niche and corner. Some of it was never cleared out to the end of the War.[1]

Comparable comments were voiced after the fall of France in 1940. The most famous and articulate British military commentator of the first half of the twentieth century, Basil Liddell Hart, who had had a hand in drafting Lloyd George's memoirs, performed similar sleights of hand in his own. He attributed the German victory to the British army's love of the horse and its

consequent resistance to mechanisation. As unofficial adviser to Leslie
Hore-Belisha, the secretary of state for war, between 1937 and 1938, he had
seen senior officers who, in his view, had been too old and too reluctant to
adapt its organisation to tactical innovation. Although he regularly accused
British army officers of resisting mental activity of any kind, that was not
the charge Liddell Hart levelled at those on the general staff in 1937–8.
Rather, they had (in words used by Percy Hobart to Liddell Hart) 'acute
attorney minds', which they employed 'in dissecting and belittling [Liddell
Hart's] proposals – and in demonstrating that nothing should be done'.[2]
Liddell Hart portrayed these veterans of the First World War as the victims
of their own victory in 1918. His conclusion was both dogmatic and
simplistic: 'History shows that, as a rule, armies learn from defeat but not
from victory – that it is the losing side which turns to account the lessons of
a war, while the victors become dangerously complacent.'[3]

The inter-war army to which Liddell Hart turned to back up this statement was that not of Britain but of France. In the 1920s the French army
thought very hard about the lessons of the First World War, and it would
have been even worse prepared for what befell it in 1940 if it had not done
so. Its conclusions – that firepower kills; that the defensive dominates; and
that a future European war would be a long one in which economic mobilisation would be vital – were perfectly appropriate readings not only of that
war but also of the next one. The Second World War was long, and it was
won by the application of resources and superior firepower. But these
lessons did not prove right for the opening campaigns of 1939–41. The risk
in Liddell Hart's precept is that it becomes an excuse for just the sort of
mental laziness he professed to abhor.

For, a far greater danger than that of becoming trapped in the past is the
failure to think about the past at all. Lloyd George compared

> the experience acquired by a doctor in the course of his practice with
> that of the professional soldier. A physician fights a series of battles
> with the enemy every day of his professional life. That experience adds
> to his mastery of the art to which he has dedicated his abilities.

By contrast, 'A soldier may spend his lifetime in barracks or colleges without
a day's actual experience of the realities with which he will have to contend
if war breaks out.' He went on to conclude: 'War is an art, proficiency in
which depends more on experience than on study, and more on natural aptitude and judgment than on either.'[4] Lloyd George was sound in his opening
premise but wrong in his final precept. Reliance on natural aptitude was the
path to amateurism, not to professionalism. The fact that armies might draw
false conclusions from previous wars did not mean that the business of
learning lessons was itself vicious. It meant, rather, that that process had to
be undertaken with particular rigour. Soldiers cannot escape the problem
which the analogy with the physician presents. They may not wage war every

day, but that is precisely why in peacetime they have to study it. 'More than most professions', Douglas MacArthur opined when US Army Chief of Staff in 1935,

> the military is forced to depend upon intelligent interpretation of the past for signposts charting the future. Devoid of opportunity, in peace, for self-instruction through actual practice of his profession, the soldier makes maximum use of historical record in assuring the readiness of himself and his command to function efficiently in emergency. The facts derived from historical analysis he applies to conditions of the present and the proximate future, thus developing a synthesis of appropriate method, organization and doctrine.[5]

The British army has too often taken counsel from Lloyd George rather than from Douglas MacArthur. All Liddell Hart's work was devoted to the use of military history to illuminate present problems: what his career highlights is not the inutility of learning lessons from the immediate past but the difficulty of doing it with rigour and objectivity. Throughout the twentieth century the British army was perhaps the most cavalier of that of any major power in its approach to the study of military history. Unlike that of Prussia or France, it did not establish a historical section in its general staff; unlike that of the United States, it did not set about the systematic acquisition of materials related to wars in which it had not itself been engaged or of translating accounts written in languages other than English. In 1972 it reduced the commissioning course at the Royal Military Academy Sandhurst from two years to one, and for the rest of the century the place of military history in the curriculum was marginal.[6] Military history played little role in the education of staff officers until the establishment of the Higher Command and Staff Course at the Staff College in 1988. Unlike other armies, the British army did not employ historians in theatre when on operations, and as a result it had no clear procedure with regard to learning lessons or to disseminating them. The Army Historical Branch in the Ministry of Defence was not central to thinking about the army's future and seemed to spend too much of its time servicing the needs of in-house struggles over issues of precedence and tradition.

Paradox lies at the heart of the British army's relationship to its own history. To an extent unparalleled elsewhere it enjoys the benefits of continuity. The German army found itself the servant of a discredited regime in the twentieth century not once, but twice. The French army was split by the collapse of the Third Republic in 1940. It did not begin to reconcile the political consequences until the 1960s, and even then it had to do so against the background of colonial withdrawal. That deepened, rather than resolved, the anguish of defeat in 1940. By contrast, the British army has not had to assimilate political upheaval since the 'Glorious Revolution' of 1688–9. Like the French, and like the Americans in Vietnam, it has suffered defeats, but it has done so by recasting them as victories – from Mons in 1914 to Dunkirk in 1940. Even colonial withdrawal, so damaging to the unity and cohesion of every other imperial army, was packaged as a success story. By avoiding profound shocks, the army has created institutional continuity. But there has been a price. Its view of history is selective and self-serving, more geared to the preservation of that sense of continuity than it is cognisant of reality.

One very good reason for this lack of introspection has been entirely functional. The twentieth-century British army was simply too busy to stop and consider why it was where it was and whither it was directed. It never went from one major European war to another without passing through a small war – or wars – on the way. It was subject to almost continuous conflict in many areas of the globe, and often in more than one place at the same time. This practice was the enemy of theory. Armies on the continent of Europe had a clearer sense of the likely terrain over which they might be called on to operate. Moreover, their enemies would – most probably – be their nearest neighbours. From this doctrine could flow, and as a result doctrine's principal servant – military history – had a role. By contrast, the British army could not anticipate with any accuracy against whom it might next fight or where it might next be deployed. This was the reason why it could not develop, and in the eyes of many should not have developed, doctrine. In 1900 G.F.R. Henderson was the army's principal advocate for the professional study of military history. An infantryman, a history scholar of St John's College, Oxford, and an instructor first at Sandhurst and then at the Staff College, he declared in an entry in the 1902 edition of *Encyclopaedia Britannica* that, 'in soldiering there is more to be learned from the history of great campaigns than from the manoeuvres of the training-ground'.[7] However, two years earlier, he had also written that,

> It is useless to anticipate in what quarter of the globe our troops may be next employed as to guess at the tactics, the armament and even the colour ... of our next enemy. Each new expedition demands special equipment, special methods of supply and special tactical devices, and sometimes special armament.[8]

In Henderson's eyes, therefore, history and doctrine could not be as integrated for the British army as for other armies. Military history was a prime tool in the education of a soldier, but its relationship to doctrine was unclear, since doctrine itself could be a straitjacket which could compromise flexibility. It could prepare the army to face the wrong enemy at the wrong time and in the wrong place. Studying the campaigns of Napoleon might not help in a war against the Zulus; familiarity with the geography of the Shenandoah valley (the complaint of Henderson's pupils who had been nurtured on his life of Stonewall Jackson) might not aid the commander on the northwest frontier of India. This argument was not just familiar to the army of the British empire; it proved as relevant to the army which withdrew from that empire in the 1950s and 1960s, and then found itself re-engaged outside Europe after the end of the Cold War.

These conditions have been so pervasive and so persistent that, for some commentators, they have become the web and woof of the army itself. In 1932 Basil Liddell Hart enunciated what he called 'the British way in warfare'. He argued that Britain's historic practice had been to use its seapower to place small expeditionary forces on the continent of Europe, while leaving the bulk of the fighting in a major war to its allies.[9] Liddell Hart's characterisation of the past was highly selective. However, his purpose was not to influence historians but to shape current and future policy. The lesson that he had learned from the First World War was that Britain should not commit a mass army to the continent of Europe again. Although Britain could not be faithful to his precepts in 1939 or 1944, it tried to be after the Second World War was over. Between 1945 and 1950, as Paul Cornish's chapter in this book makes clear, the British army did not favour a continental commitment in the sense that it did after the formation of NATO. The British Army of the Rhine was an occupation force only, designed to keep the Germans down, not to keep the Russians out. Montgomery, as Chief of the Imperial General Staff, accepted that Britain's principal arm in a future major war would be the Royal Air Force, whose bombers were an updated and recast version of British maritime power.

Liddell Hart's formulation can be seen as evidence that strategy is the product of culture, nationally variable not universally applicable.[10] But if we are to conclude that the British army's approach to the study of war, and to the learning of lessons from military history, is also culturally determined, we need to be clear what the components of that culture have been – and still are. In the eyes of David French and of Colin McInnes, both writing in this volume, it is that of 'muddling through', and even making a virtue of it. In the eyes of the German army, its principal opponent in two world wars, the British army had two cultural characteristics in October 1939. The first was that it accepted 'losses and setbacks with equanimity'. On the basis of its performance in the First World War, that was a fair assessment. But a year later both the prime minister and his generals were worried by the growing evidence to the contrary: in France and Singapore British soldiers surrendered too readily.[11] The second characteristic which the Germans identified was that the army's leadership was 'schematic and slow. Operational deftness and spirited handling of tactical maneuvers are not to be expected'.[12] But from the late 1980s the British army's doctrine was centred on what it called 'manoeuvrism', a concept which had less to do with manoeuvre traditionally defined and more to do with tempo. The object was to act with sufficient flexibility and responsiveness, in order to get inside the enemy's decision-making loop. In other words neither characteristic, however culturally conditioned, seemed to command continuity throughout the twentieth century.

Cultural concepts of strategy may be no more helpful, and no more subtle, than national stereotypes. Liddell Hart's 'British way in warfare' was an idea about the conduct of major war. It addressed the issue of whether, in the event of a war with a European power, the navy or the army should be Britain's weapon of choice, and came down in favour of the former. Two years before its publication, the 1930 edition of the army's *Field Service Regulations*, as quoted by David French in his essay on the inter-war army for this volume, stated as its opening premise that, 'the British Empire is confronted with problems peculiarly its own'.[13] The general staff's cultural determinant therefore was not the relationship between maritime power (or – later – air power) and military power, but that between major war on the continent of Europe and the small wars of empire. If the British army ever embraced a 'British way in warfare' (a proposition more proveable in hindsight than through contemporary evidence), it must have included colonial campaigning, which constituted its principal ongoing commitment and for whose conduct it was largely structured.[14]

Therefore, the prime cultural influence on the British army has been one not so much of geography as of scale. It has had to be capable of fighting a war of the first magnitude while simultaneously being ready to engage in wars of smaller dimensions. This was the philosophy which the general staff enunciated in the inter-war period. It was not the philosophy of the Cold War, but it was the practice. The army was quartered in north Germany, readying itself for a possible Soviet invasion, but it was shot at in Malaya, Kenya, Cyprus, Aden and the Falklands. With the end of the Cold War, it became the philosophy once again. In 1993 the Conservative government used Britain's membership of the Security Council of the United Nations to redefine the global reach of its defence policy. When Labour came to power in 1997, loyalty to the 'special relationship' with the United States had similar – if further reaching – effects. Britain's army was engaged in active operations almost without a break, but few of those operations could be defined as wars. However, it continued to embrace doctrinal principles founded on the presumption of major war. Manoeuvre warfare, the leading idea of the 1980s, formulated in the context of the Cold War for the conduct of a corps level battle against a Soviet invasion of the inner German border, remained the central concept for the army after 1990. But now its operations involved much smaller formations, very often engaged in peace support and peace enforcement.

Throughout the twentieth century, therefore, the army has made major war the conditioning factor in shaping its doctrine. 'The instructions laid down herein', the *Field Service Regulations* of 1929 declared, 'cover a war of the first magnitude, but are to be modified in their application to other forms of war.'[15] By and large that has meant war against a European opponent, even if that opponent might not be confronted within Europe itself. For most of the nineteenth century, Britain's most likely enemy – apart from France – was Russia. An Anglo-Russian war would have been fought in

India. And India did not cease to be one of Britain's most probable battle-fields after the entente with Russia in 1907. It was the base for Britain's army in the war against Japan, and as a result in 1941–5 Britain fought a major war not on India's northwest frontier, but on its northeastern. The British army after 1945 may not immediately have prepared for major war in Europe, but it still prepared for major war against Russia. Its focus, as Paul Cornish says in this book, was on the Middle East. In 1950 it got its continental war, but the war was in Korea, and the continent and the enemy (China) were Asiatic.

By taking major, not minor, war as its gold standard, the army reckoned that it was simplifying the complexities of life. Shifting down the scale seemed to be an easier task than shifting up it. The army first used and then rejected the title 'low intensity operations', a phrase made famous in 1971 by one of its more idiosyncratic generals, Frank Kitson, who developed his thinking on the basis of his own experience in Kenya fighting the Mau Mau and went on to apply it in Northern Ireland.[16] Those who disliked the term did so because men under fire are in a situation as dangerous – as intense – as those fighting a major war. In other words the distinction between high intensity warfare and low intensity operations might make sense in the context of strategy but made far less sense at the level of minor tactics. By the same token minor tactics acquired greater salience in low intensity operations.

Those hardy perennials of tactical debate, the relationship between fire and movement, therefore provided an element of continuity across the spectrum of conflict. But that did not mean that tactics were constant between major war and minor war, or between different geographical environments. The challenge for armies at the beginning of the twenty-first century was to think through the tactical adaptations imposed by fighting insurgents in urban areas. This was a form of war very different from the air-land battle which corps commanders planned to fight with armoured divisions in northern Europe in the 1980s. Scale and restraint in the use of force were not the only distinctions; climate and terrain were as significant – if not more so. The Second World War provides an instructive comparison. By 1942 the British army had adjusted to the tactics of mechanised war in North Africa, where the desert provided clear fields of fire and wide spaces for manoeuvre. But in 1943–4, it had to adapt again – and in three different ways. Mountain warfare prevailed in Italy. Armour found itself caught in traffic jams on narrow roads. Second, in the battle for the defence of India's northeastern frontier, the army had to learn to fight in the jungle, as Daniel Marston explains in this volume, and then applied the precepts in the Burma campaign. Third, in the breakout after D-Day, the *bocage* and enclosed landscape of Normandy required the army to develop small group cooperation between tanks and infantry.[17]

Tactical adaptation of this sort explains one important cultural consequence of the lesson-learning process in the British army. Front-line units initiated change, responding pragmatically to the circumstances that confronted them in the field. Daniel Marston gives one striking example: the 2nd battalion of the Argyll and Sutherland Highlanders, alone of the components of 12th Indian Brigade in Malaya in 1942, set about learning the techniques of jungle warfare. Similarly, the Gurkhas led the way in the rediscovery of jungle warfare after 1945. Minor wars in the twentieth century, like colonial garrisoning in the nineteenth, favoured the infantry battalion – not the brigade or division, let alone the corps – as the key command level. Innovation from the 'bottom up', rather than from the 'top down', was both a symptom and a cause of regimental autonomy.

At its best the regiment fostered change; at its worst it prevented the dissemination of best practice throughout the army. It gave commanding officers the power to reject good ideas, as well as to embrace them. It meant that lessons learned by one battalion might not be passed on to their successors or be forgotten by that same battalion when deployed in a different theatre on different duties. It made the institutional memory of the army, at least in regards to tactics, very fragile.

The conduct of low intensity operations reinforces this point. The minting of generic terms for minor conflicts – others include counter-insurgency war and operations other than war – does not in itself mean that those forms of war were subject to a common doctrine. In 1967 Colonel Julian Paget published *Counter-insurgency Campaigning*, a book endorsed with a foreword by the Director of Army Training. Paget used the campaigns in Malaya, Kenya and Cyprus to draw out general principles for such operations. His book created the impression that the British army had developed universally applicable techniques for dealing with guerrillas, the core of which had been laid in Malaya by Field Marshal (as he became) Sir Gerald Templer, and applied successfully elsewhere during Britain's withdrawal from its colonies. In practice Paget was guilty of rationalising after the event. The year in which he wrote, 1967, was the concluding phase of Britain's imperial retreat: only the withdrawal from Aden remained, and that was not the most distinguished episode in the saga. *Counter-insurgency Campaigning* suggests a parallel with *Small Wars: Their Principles and Practice*, by C.E. Callwell, first published in 1896 but best known in its 1906 edition. Like Paget, Callwell synthesised the lessons learned from colonial warfare just as an era, in this case that of colonial conquest and settlement, was ending. The effect of *Small Wars*, like the effect of *Counter-insurgency Campaigning*, was to suggest that the British army conducted its minor conflicts with greater consistency than was actually the case. In each of Paget's examples, Malaya, Kenya and Cyprus, although some expertise was imported from one area to another, the commanders on the ground developed their own responses independently and without reference to counter-insurgency campaigns being waged elsewhere.[18]

If Malaya had an immediate impact, it did so in thinking about the problems of confronting communism within Europe, not in thinking about handling the withdrawal from empire. Between 1945 and 1950, as Paul Cornish's chapter shows, the army saw the principal defences of Europe to be ideological and political, not military. Therefore the greatest Soviet threats were subversion and sabotage, not direct attack. The Russians would come only when the internal fabric of the state had been corroded from within. The job of the British Army of the Rhine was to counter civil disobedience. This element remained central to ideas about the defence of the West after 1950. In 1957 an official Ministry of Defence publication, which examined the Malayan campaign, did so within the context of guerrilla war, setting it alongside partisan operations in Europe in the Second World War. This was pre-1950 thinking in reverse: the best responses to a Soviet take-over of Europe also lay in subversion and sabotage. The British army was therefore interested less in its own counter-insurgent successes than in the methods of its opponents, the insurgents. These were an example for emulation in the case of another war in Europe.[19]

Malaya became a model of counter-insurgency after the events, not during them. Indeed, even in 1968 Brigadier R.C.P. Jeffries declared that lessons learned from other theatres were inappropriate in Aden. David Benest, who cites Jeffries in his chapter in this volume, goes on to show how in the following year, 1969, the British army became the victim of its own lessons-learned process when it was first deployed on the streets of Northern Ireland. It had to adapt in the middle of a conflict by learning fresh lessons, just as it had done in the two world wars. Techniques acquired in Malaya and Aden were not necessarily transferable to an insurgency fought within the United Kingdom under the full glare of the press. 'Hearts and minds' in the 1950s and 1960s had been accompanied by coercion, collective punishment and compulsory resettlement. *Land Operations: Volume III – Counter-revolutionary Operations*, published in November 1969, did give instructions for the suppression of riots and the control of urban movement. But its illustrations showed troops in Aden, not in Europe, and its instructions regarding crowd dispersal read:

> Warn the crowd by all available means that effective fire will be opened unless the crowd disperses at once. This can be done by a call on a bugle, followed by the display of banners showing the necessary warning in the vernacular.[20]

These instructions do more than reveal the assumption that the language of the crowd would not be English. They also show that counter-insurgency doctrines gave salience to methods which were inappropriate in major conflicts, a point whose tragic consequences are highlighted by David Benest. The British army suffered heavier average casualties in the more benign conditions of Northern Ireland than it did in the 'hotter' and more overtly hostile circumstances of Iraq in 2003–4. Advancing German infantry in the Second World War was not alerted to the fact that they were about to be fired upon. In 'operations other than war' the principle of tactical surprise, so central to fighting a 'hot' war, was therefore deemed secondary to the observance of rules of engagement. Armoured fighting vehicles were to be kept in reserve, and designed to overawe not to destroy. By the same token artillery, the key component of the combined arms battle of the two world wars, had no obvious function in suppressing insurgencies – or at least not those when discrimination in fire effects was deemed mandatory. Ultimately the army was deployed as an aid to the civil power, not in its own right. The doctrine for its use was predicated more on the legal authority under which it acted than on its military capability.

The need to use force with restraint was the severest challenge at the level of minor tactics in low intensity conflict. The fear that it would undermine the 'warrior' ethos of the American soldier underpinned the United States Army's assertion that it did not 'do' peacekeeping. The British press in the second Gulf War, with typical chauvinism, suggested that low intensity operations, in which it saw the British army as particularly expert, required greater skill and finesse than did war-fighting proper. At the level of the individual soldier, out on patrol, unsure as to who was the enemy, and having to balance his rules of engagement with his own protection, this might be fair comment. But away from the level of the individual soldier, the shift from low intensity to high intensity operations proved much more difficult than the shift from high to low.

This point can best be illustrated at the level that for armies in the mid-twentieth century would have seemed the most abstract of all, the laws of war. German soldiers may have been charged with war crimes after the Second World War, but for the British army of the 1950s the laws of war were not a major factor in shaping doctrine even in the context of counter-insurgency. In Kenya the British army did not embrace the principle of minimum force in suppressing Mau Mau. In April 1956 alone 1,086 suspects were executed, and during the insurgency as a whole 78,000 were detained, and 402 prisoners died in captivity. When the insurgents fled to the forests, the Royal Air Force bombed them, despite its inability to identify its targets.[21] As journalists reported from the flash points of the world in the 1960s, such procedures became increasingly unacceptable, even in the Middle East. The deployment of the army in Northern Ireland, and especially the aftermath of 'Bloody Sunday' in Londonderry on 30 January 1972, consolidated the principle of minimum force, and the importance of rules of engagement. 'Everything done by a government and its agents in combating insurgency', Frank Kitson wrote in 1977, 'must be legal.'[22] 'Small wars' therefore were the context in which the British army learned to apply the principles of discrimination and proportionality in conflict. But by the end of the twentieth century those principles had become central to the conduct of all conflicts, not just to ones of low intensity. On D-Day, in the Second World War, the allies killed more French civilians – non-combatants, and those they had come to liberate – than Germans. When the British and

American armies invaded Iraq in April 2003, they were committed to a major operation of war, but their doctrines were shaped by the principles of non-combatant immunity and discriminate use of firepower. Lessons learned in small wars had percolated up to bigger ones, not least as a result of media pressure and public observation.

That could present those engaged on the ground with tough decisions, especially in urban areas where combatants and non-combatants might be in the same building. But even tougher at an institutional level were other aspects of the gear-change to major war. Most complex of all was that of logistics. One obvious legacy of the British army's colonial role has been its reach; it knows better than most that sustaining an army operating at half a hemisphere's distance requires the ability to get supplies to it. David French argues that it was precisely this lesson from colonial warfare which stood the army in such good stead in the Second World War. But major war also required a bigger army, and that in turn needed more weapons and more equipment – indeed more of everything. If Liddell Hart's 'British way in warfare' had a positive legacy, it was that Britain could feed a small expeditionary force from the sea. But its negative consequence was that that army was small and the industrial base geared to its equipment was also small.

A doctrine for major war is not the same as the capacity for major war. Small wars need fewer shells; they do not require arms manufacturers to have a 'surge' capacity; they do not require British industry to convert to war production, a process which took two years in the First World War. This is not to say that small wars favour conservatism in weapons procurement. Colonial conflict encouraged the British army to develop the use of the machine gun as a force multiplier. As Edward Spiers shows, the Royal Artillery adroitly used its humiliation at the hands of the Boers to make the case for quick-firing artillery. But small wars, especially those fought at a distance from the United Kingdom, put a premium on lightness, to enable rapid deployment. David French cites several examples of inter-war procurement decisions, conditioned by the problems of re-supply, which resulted in sub-optimal equipment for major war. The British army's pursuit of expeditionary warfare after 1990 confronted it with similar challenges. Tanks designed for temperate climates and for major armoured battles were not necessarily suitable for desert warfare. But by reconfiguring for 'lightness', the army was in danger of elevating the principle of mobility over those of protection and of fire effect.

In major war more men are needed at the outset and more men get killed over its course. In 1914–16 and again in 1939–41 the British army grew tenfold. The short-term consequences were chaotic: men with inadequate equipment were led by officers who had been promoted beyond their abilities or expectations. In both cases the army had given serious thought to a possible war in Europe, but it was not prepared for such a war when it broke out. In 1914, the secretary of state for war, Lord Kitchener, thought the army would not be ready for major operations on the continent until 1917, even if it actually found itself committed to such operations before that. In 1940, the army was knocked off the continent, not least because it was in disarray as a result of over-rapid and ill-thought-through expansion, and it was not ready to renew the fight in northern Europe until 1944.

This is not to say that the British army did not possess mobilisable reserves in the event of major war, but it proved remarkably reluctant to embrace the lesson that it might need them. The biggest such reserve, at least until 1947, was the Indian army. Before 1914 the Indian army was, in contemporary jargon, 'double-hatted'. It existed, first of all, to enable Britain to hold on to what it had, to deal with domestic disorder within the sub-continent. To fulfil that role it did not need levels of armament or organisational structures comparable with the most advanced armies of the day. Indeed it might be a positive disadvantage if it did: the biggest threat to British rule in India had been internal, the mutiny of the Bengal army in 1857. If that happened again, British troops would depend on superior armament to offset their inferior numbers in order to re-establish order. But the Indian army had a second function, to defend India against the external threat of the Russian army, which it was feared would bear down on India through Afghanistan. If that happened, it would need to be armed and organised to the latest standards.

In 1914 India sent expeditionary forces to Europe, Egypt, Mesopotamia and East Africa. It was therefore treated as a reserve in the event of major war. All four forces confronted significant problems – even if for different reasons. Not until 1917–18 did the Indian army prove effective in the field, and then not in Europe but in the Middle East in the war against Ottoman Turkey. In 1941–2, Indian formations were again committed to major war, against Japan, and again suffered initial defeats. But by 1944–5, as Daniel Marston shows, the Indian army was a key player in the developing techniques of jungle warfare.

Britain's other reserve was the Territorial Army. However, as with the Indian army, the regular British army proved consistently ambivalent about its place in the event of major war. In 1900 the army had to use its predecessors, the Militia and Volunteers, to meet the manpower crisis generated by the South African War. But the principle of expansion in the event of another big war was not among the lessons which the army digested in the South African War's aftermath, and which Edward Spiers discusses in his chapter. When Richard Burdon Haldane, as secretary of state for war, reorganised the Militia and Volunteers to create the Territorial Army in 1907–8, the new organisation's declared purpose was home defence: it was designed to release the regular army to go overseas, but it was not itself prepared to go abroad to reinforce that army. In the event it did so, but its role in the First World War was overshadowed by Kitchener's decision not to rely on it but to create wholly 'New Armies' in 1914.

In 1939–40 the army did turn to the Territorials, who provided half of the British Expeditionary Force sent to France (and more if labouring and lines of communication troops are included).[23] From at least 1936 the general staff had realised that only the Territorial Army could provide a reserve for the regular army in the event of European war, and had hoped to have twice as many fully equipped Territorial divisions as regular. But right up until March 1939 the Territorials' primary roles remained home defence – the provision of anti-aircraft batteries and the maintenance of domestic order in the event of German air attack. Then the government decided to double their strength, a clear indication that it too now saw them as a reserve for continental war. This was far too late: neither their equipment levels nor their training were commensurate with the task loaded on them in 1940.[24] Nonetheless, in 1950, ten years and a world war later, confronted with the possibility of sending an army to war in Europe, the army looked again to its Territorials, although it would take between three and six months to mobilise them.

This was one relationship which the regular army never resolved in the twentieth century, despite the lessons of war. The philosophy of 1939–40 was that of one army: the divisions of the British Expeditionary Force mixed regulars and Territorials, and the latter were told to drop the distinctive 'T' from their uniforms. The Territorials saw that as an insult to their own institution. The regulars were just as bad, refusing even to don the uniforms of the Territorial units to which they were posted as commanding officers or adjutants. As Colonel E.J. King wrote in 1943: 'The less intelligent type of staff officer, in the sacred name of discipline, lost no opportunity of shewing his contempt in the most offensive and provocative manner for the TA and all its ways.'[25] In Italy, on 28 May 1944, Gordon Simpson, a yeoman who was to be awarded the DSO and bar, and who at one stage would be acting commander of 26th Armoured Brigade (and therefore of two regular cavalry regiments as well as his own), wondered how secure was his command of his own regiment, 2nd Lothians and Border Horse. He confided to his diary:

> I still have the ever-present spectre of some suitable Lt. Col. arriving from way back, either a loathsome little tick from the Tank Corps or else an equally horrible cast-off from some cavalry regiment – we've had plenty of experience of them this war.[26]

Nor was this tension simply a matter of perceptions about training and competence; it was also a battle for resources, as became clear in 1965, when two regular officers of immense distinction, Sir John Hackett and Michael Carver, proposed that the sole function of the Territorial Army should be to bring the regular army from a peacetime to a war footing. The old and bold of the Territorial Army fought back, but by 2003 the regulars' solution, that of 'one army', drawing on individual reservists for full-time service with regular units, had undermined the unit cohesion of the Territorial Army. Lacking a clear identity in regional terms, it struggled to recruit; treated as second-class citizens in terms of equipment and training time, its members were deployed on operations with insufficient preparation. Moreover, the concept applied the principles of major war to minor conflicts, making the regular army dependent on the Territorial Army even when no national emergency existed.

The succession of defeats suffered by the army between 1940 and 1942 demonstrated that the shift from minor war to major war was not just an intellectual step-change internal to the army. It was also a political decision. David French argues that the army got it more right than wrong in the years 1919–39: it embraced the lessons of the First World War long before they were specifically addressed by the Kirke report of 1932, and it knew that, even while it was fighting in Iraq or on the northwest frontier of India, it needed to measure itself against European opponents.[27] But that was not the same as persuading its political masters to fund and equip the army in line with the philosophy. Lessons, once learned, could not be converted into effective reforms without political support.

Most importantly the decision to go to war was one taken by politicians not soldiers. Rarely in the twentieth century did the army's lesson-learning process converge with that decision. As Edward Spiers stresses, the army after the South African War reformed itself the better to fight an imperial war. This was not because it was a captive of the past, for all Lloyd George's later strictures, but because that conformed with government policy and with the resources that the army was allocated. Even when the cabinet eventually decided to enter the First World War, it failed to register the consequences for the army – massive expansion and conscription, the obvious lessons of earlier continental wars: instead it thought it would wage a maritime war. In 1939, the army found itself going to war for Poland, not for France or the Low Countries as it had expected, and doing so with an army that bore little relationship to what it had said it would need if there was war. In 1982, as Simon Ball argues, the lessons the army had been learning were those deemed appropriate to NATO's central front. The army was geared to the defence of Europe but was then used as the 'projectile' of the navy in the south Atlantic. A sailor, Sir Terence Lewin, was Chief of the Defence Staff, and the First Sea Lord, Sir Henry Leach, managed to get the prime minister's ear. If Lord Carver, a soldier and continentalist, had still been Chief of the Defence Staff, Mrs Thatcher might well have been cautioned not to act.

The first Iraq War of 1991 bore comparison with the Falklands war: the army was focused on Europe but found itself fighting in very different conditions, both in terms of terrain and weather. It used methods which it had developed for the possible conventional battle in Europe, but in this case the effects, according to Colin McInnes, were longer-lasting: they became the basis for a paradigm shift, and confirmed the belief that preparing for major war enabled the army to fight any war. What that ignored was that preparing for major war had actually made the army wary about fighting at all. Then and since, senior soldiers have tended to predict that casualties will be greater than in fact they have proved to be. The prevailing image of Britain's wars after the end of the Cold War, whether in Kosovo in 1999 or Iraq in 2003, was that they were waged at the behest of politicians in circumstances and under conditions which the army did not regard as propitious.

The army therefore had to accommodate the consequences of its own success. In the late twentieth century, victory created its own difficulties in learning lessons. These were not the vices of complacency or arrogance, those highlighted by Liddell Hart, so much as the loss of political leverage which victory conferred. Defeat creates its own unanswerable case for reform, as much in the military as outside it. The army of 1918 and the army of 1945 were the results both of lessons learned in the early setbacks of the two world wars and of the government's readiness to fund and implement change. As the army found in the 1990s and even after the second Iraq War, the case for resources was not one which had much political leverage when the army seemed to be doing very well on what it had. In 2004, with the army still deployed on operations in Iraq, it identified changes which it felt to be necessary in terms of its capabilities, but it could only find the funding by cutting the total number of infantry battalions from forty to thirty-six.

Nor were the political problems confined to the army's relationship with the government. It could also be a matter of inter-service manipulation. Over the course of the twentieth century the army found it harder to make its case for funds in peacetime than did the Royal Navy or the Royal Air Force. In part that was a reflection of the procurement time for large items of sophisticated equipment, like ships, aircraft and missiles. But in part it was also a reflection of the power of deterrence as a peacetime tool, whether embodied in the fleet, the bomber or the nuclear weapon. One lesson that the army did learn therefore was the need to play off inter-service issues in one of two ways. Either it could divide and rule: in other words it could form an alliance with the one of the other two services at the expense of the third. This is what it did in the Cold War, and especially in the era of the Nott reforms of 1981–2, when the Royal Navy became the principal casualty of the other two services' focus on the continent of Europe. Or it could conform to the grain of strategic orthodoxy but then reinterpret it according to its own best interests. The argument that the air defence of Britain began on the continent – used both before the Second World War and after – is a case in point. The army accepted that air power was Britain's principal weapon in the event of European conflict but then recast it in its own interest.

Inter-service jockeying for position worked against joint operations. Counter-insurgency warfare was often joint at the tactical level: David Benest provides plenty of evidence of the army's reliance on airpower. But that did not lead to 'jointery' in institutional terms. Simon Ball's chapter on the Falklands reveals how little cooperation and mutual understanding had developed between the army and the navy despite their interdependence. Joint command, embodied in the creation of the Permanent Joint Headquarters at Northwood, was not, he argues, a lesson learned from the Falklands but a consequence of the end of the Cold War, and even then bitterly contested. The NATO doctrines of the 1980s had emphasised the integration of the air and land components in the so-called 'air-land battle', but even in the first Iraq War of 1990–1 – Colin McInnes argues – this ideal was a long way off realisation.[28]

For the army the development of joint command created challenges to the lesson-learning process. First, its advent rested on the presumption of expeditionary warfare on a global scale, a threat to an army configured for the use of heavy armour in battles in Europe. The army could accept that it was unlikely to use its tanks within Europe, but it could still envisage fighting major war outside Europe, just as the army in India before the First World War or the army in the Middle East after the Second World War had done. Second, joint doctrine discounted its own heritage, which was predicated on a single service history and single service hierarchy. There were still issues to do with land warfare – just as there were issues to do with war in the air or at sea – that seemed best understood and resolved at the single-service level. Joint warfare had the tendency to plane rough surfaces smooth so that common, inter-service and inter-operable methods could be established. But the presumption behind the fashionable ideas of the 1990s and early 2000s, and specifically 'the revolution in military affairs', network-centric warfare and effects-based operations, could easily be that history was a false friend, even if those who took part in the debates which they generated were ready both to use and to abuse military history.

In the event what was striking about the changes in defence after the end of the Cold War was how little inter-service squabbling they produced – at least in public. The key to the learning of lessons has been the general staff, established in 1904–6. The general staff was in theory responsible for sifting the ideas from below and from the examples set by the armies of other nations; it could even have ideas of its own. From this process emerged common concepts, which might prove flawed but which at least provided unifying themes. By Britain's own standards, the inter-war general staff was a reasonably cohesive body with a clear sense of belonging to the army not to a collection of regiments. That was because it owned a coherent philosophy predicated on the possibility of major war. But, like its pre-1914 predecessor, it could not impose that philosophy on the army as a whole. The post-1945 British army lost even that coherence, principally because the twin pressures of the Cold War and colonial withdrawal created two different sets of interests, one focused on Europe and armoured divisions, and the other revolving round points east of Suez and infantry battalions. As Simon Ball highlights, to those from the latter school those from the former were Whitehall warriors cut off from the realities of soldiering. The threat of major war made them unsympathetic to lessons derived from lesser conflicts. This division was played out in the story of regimental amalgamations and disbandments which accompanied every major defence review between 1957 and 2004. The lessons became ones not of tactics but of institutional survival, and even senior staff officers gave their loyalties to their regiments over the army as a whole.

The fact that things began to change in the 1980s, as Colin McInnes describes, was due to the dominance of the British Army of the Rhine. Colonial withdrawal having been completed, the army could concentrate on the continental commitment. In 1981 two former officers of the Royal Artillery concluded a major study of the army's thinking about war in the first half of the twentieth century on a sombre note:

> The British army is not an institution able to express views, and to propose decisions on professional grounds alone, allowing the politicians both the right and the responsibility of disposal. To that extent, despite its achievements and reforms since 1906, the Army remains what it was then, *sans* doctrine and an unprofessional coalition of arms and services.[29]

But even as they were writing things were changing, under the guidance of Sir Nigel Bagnall, commander of 1 British Corps in Germany. He stressed the corps battle and the operational level of war. In 1988 the Higher Command and Staff Course was established at the Staff College, and in 1989 the first-ever British military operational doctrine was published.[30]

However, all this happened within the context of Europe. After the end of the Cold War, doctrine developed for one region was stretched and diluted to do duty in others. At the same time, the need to think and plan in conjunction with the other two services in environments that were now entirely joint increased, not diminished, the need for coherence and for clear thinking about the past and about the appropriate processes for learning lessons. Joint organisations made it even more tempting to choose examples or solutions that rested on selective readings of the past – that confirmed the joint orthodoxy or that rested on a common denominator, even when it was the lowest.

Ultimately history is about asking questions before it is about finding answers. All the contributors to this volume in their various ways make a related point. The source of the lessons is often not what subsequent critics assume it to have been. Edward Spiers points out that the origins of the reforms, which Lord Roberts backed when he became the army's commander-in-chief in 1900, lay not in South Africa, the war from which he had just come, but in his experience in India and in currents of thought present in the 1890s. Moreover, the lessons of war, even when conscientiously studied, can be ambiguous: both sides in the debate over the future role of the cavalry could cite the South African War in support. David French shows the ambiguity of the First World War for the British army in the inter-war years: a fruitful source of lessons had to be handled with care because of the public horror at any suggestion that there might be another war of trenches, blood and mud. Of course the army did learn lessons from the First World War which it applied in the Second: the battle of El Alamein itself, the first and last independent British victory over German forces, was the direct descendant of the western front offensives of late 1918. And in the winter of 1944–5 the army fought battles in northwest Europe that would have been all too familiar to veterans of the Somme or Passchendaele.

What is most striking about Paul Cornish's chapter is the discontinuity between the Second World War and its immediate aftermath: the army did not embrace a continental commitment after 1945 although it had done so in 1944. It remained a mass army but it anticipated major war occurring elsewhere in the world, as indeed it did in 1950. When it did commit itself to Europe, it built on the legacy of 1944–5, but that was not preordained. Daniel Marston demonstrates the difficulty of holding on to lessons even if learned under the most adverse circumstances: when confronting insurgency in Malaya after a lapse of time similar to that between the two continental commitments, the British army had largely forgotten what it knew about jungle warfare. It learned the lessons again, and from the very same people who had taught it those lessons in the Second World War. By 1969 it had settled into a groove, thinking it knew how to digest the benefits of its own experience. But the danger, as David Benest shows, was that it would interpret a conflict not as it was but as those lessons led it to expect it to be. Northern Ireland revealed the dangers of learning lessons too rigidly: the failure to recognise that conclusions drawn from one environment were not transferable to another. Nor did the army prove to be quick to learn. Simon Ball highlights the principal tactical lesson from the Falklands, the infantry's need for firepower, and particularly its reliance on the general purpose machine gun, a seemingly self-evident point to the reader of any account of the battle for Goose Green or Mount Tumbledown. But the army's emphasis on manoeuvre in northern Europe led it to neglect the firefight, as – according to Colin McInnes – the first Gulf War made obvious. This was an issue which the army was still struggling to resolve a decade later, when the problems of the SA 80 rifle produced a succession of adaptations without wholly resolving the soldier's worries.

Finally there is the challenge thrown down by Colin McInnes' analysis of the lessons from the first Iraq War. That conflict seemed to confirm that major war was the gold standard, and that the thinking about the operational level of war evolved in a NATO context was transferable to other theatres. The political utility of this victory, therefore, was not just that it answered those critics who questioned the necessity for robust military capabilities after the end of the Cold War but also that it shaped how the image of future war might be configured. But because it was over so quickly, it did not resolve the fundamental conundrum as to whether the British army could sustain such operations over the long term, nor whether in engaging in peace support operations since 1991 it has become progressively less capable of fighting another major war. David Benest believes that it has. The challenge for today's army is that its philosophy is comparable with that of the inter-war army. It rests on preparation for major war but it itself is shaped and funded for small wars. It rests too on the assumption that only small wars are on the horizon. In 1982 and 1991 it assumed that a big war was more probable, but it then proved capable of switching from major European war to war elsewhere. This introduction has argued that the reverse process contains even greater challenges, even if less obvious ones from a parochial perspective. A major war would not just test the army in combat of high intensity, something other conflicts can do in any case; it would also test its capacity to recruit, train and equip a larger army over a longer period. Given the lack of political will in any democracy to confront these issues ahead of time, the proof of this pudding, of whether the army will be able to handle the shift to a major war, can only lie in the eating.

Notes

1 David Lloyd George, *War Memoirs*, 2 vols, London: Odhams, 1938, 2, p. 2038.
2 Basil Liddell Hart, *Memoirs*, 2 vols, London: Cassell, 1965, 2, p. 8.
3 Ibid., 1, p. 60.
4 Lloyd George, *War Memoirs*, 2, p. 2037.
5 Cited by Peter G. Tsouras (ed.) *The Greenhill Dictionary of Military Quotations*, London: Greenhill, 2000, p. 233.
6 Brian Bond, 'The labour of Sisyphus: educational reform at RMA Sandhurst', *Journal of the Royal United Services Institute for Defence Studies*, September 1977, pp. 38–44; Hew Strachan, 'The British army and the study of war: a personal view', *Army Quarterly*, 1981, vol. 111, pp. 134–48.
7 G.F.R. Henderson, *The Science of War*, London: Longmans, 1919, p. 49.
8 Jay Luvaas, *The Education of an Army: British Military Thought 1815–1940*, London: Cassell, 1965, p. 244; see pp. 216–47 on Henderson generally.
9 Basil Liddell Hart, *The British Way in Warfare*, London: Faber, 1932; Michael Howard, 'The British way in warfare; a reappraisal', in *The Causes of Wars*, London: Temple Smith, 1983, pp. 169–87, is the classic critique of Liddell Hart.
10 Elizabeth Kier, *Imagining War: French and British Military Doctrine between the Wars*, Princeton NJ: Princeton University Press, 1997, pp. 116–19; Colin McInnes, *Hot War, Cold War: The British Army's Way in Warfare 1945–95*, London: Brassey's, 1996, pp. 1–2.
11 Mark Connelly and Walter Miller, 'The BEF and the issue of surrender on the western front in 1940', *War in History*, 2004, vol. 11, pp. 424–41.
12 Ernest R. May, *Strange Victory: Hitler's Conquest of France*, New York: Hill and Wang, 2001, p. 256.
13 See below, p. 36.
14 See Hew Strachan, 'The British way in warfare', in David Chandler and Ian Beckett (eds) *The Oxford Illustrated History of the British Army*, Oxford: Oxford University Press, 1994, pp. 417–34; also 'The British way in warfare revisited', *Historical Journal*, 1983, vol. 26, pp. 447–61.
15 Quoted by David French; see below pp. 36–7.
16 Frank Kitson, *Low Intensity Operations: Subversion, Insurgency and Peacekeeping*, London: Faber, 1971. Kitson's introduction does more to define

what the army was fighting against, as in the subtitle of his book, than it does to define low intensity operations themselves.

17 See John Buckley, *British Armour in the Normandy Campaign*, London: Frank Cass, 2004.
18 Frank Kitson, *Bunch of Five*, London: Faber, 1977, makes this point clear at a number of levels. Kitson's thinking developed pragmatically on the ground in Kenya. He went to Malaya after that experience, not before it, and he did not synthesise it in *Low Intensity Operations* until 1969–70.
19 C.N.M. Blair, *Guerilla Warfare*, London: Ministry of Defence, 1957.
20 *Land Operations: Volume III – Counter-revolutionary Operations. Part 2 – Internal Security*, London: Ministry of Defence, 26 November 1969, p. 32, para. 101a.
21 Ian F.W. Beckett, *Modern Insurgencies and Counter-insurgencies: Guerrillas and Their Opponents since 1750*, London: Routledge, 2001, pp. 125–6.
22 Kitson, *Bunch of Five* (1987 edn), p. 289.
23 L.F. Ellis, *The War in France and Flanders 1939–1940*, London: Her Majesty's Stationery Office, 1953, p. 19.
24 Brian Bond, *British Military Policy between the two World Wars*, Oxford: Oxford University Press, 1980, pp. 27, 197–8, 221–2, 237–42, 257–8, 263–5, 288, 305, 309.
25 Quoted by Peter Dennis, *The Territorial Army 1907–1940*, London: Royal Historical Society, 1987, p. 256.
26 Gordon R. Simpson, 'A Princes Street Lancer 1936–1946', unpublished type-script, p. 36.
27 As well as his chapter in this book, see David French, 'Doctrine and organisation in the British army, 1919–32', *Historical Journal*, 2001, vol. 44, pp. 497–508.
28 See also McInnes, *Hot War, Cold War*, p. 110.
29 Shelford Bidwell and Dominick Graham, *Fire-power: British Army Weapons and Theories of War 1904–1945*, London: George Allen and Unwin, 1982, pp. 294–5.
30 McInnes, *Hot War, Cold War*, pp. 60–70, says more about this process.

1 Between the South African War and the First World War, 1902–14

Edward M. Spiers

In analysing the value of colonial operations for large-scale warfare, the experience of the South African War (1899–1902) represents an ideal starting point. Although a colonial campaign in essence, and Britain had fought colonial campaigns recurrently (and often successfully) throughout the nineteenth century, this war differed in scale, character and duration, evolving into an unprecedented challenge for the Victorian army. That army was used to short, decisive, and relatively inexpensive campaigns, but the South African War lasted thirty-two months, involved 448,000 men from Britain and the empire, and cost £230 million and almost 22,000 British and imperial dead. The defeats of Stormberg, Magersfontein and Colenso (the 'Black Week' of 10–15 December 1899), the disaster at Spion Kop (24 January 1900), and a series of costly, sometimes inconclusive engagements were seen at the time as harbingers of reform – 'we have had no end of a lesson, it will do us no end of good', wrote Kipling.[1] The reforms that followed both during the war and after, particularly the wide-ranging reforms of Richard B. Haldane as Liberal Secretary of State for War (1906–12) and the contingency and mobilisation planning of the newly formed General Staff, have been seen as producing 'incomparably the best trained, best organised and best equipped British Army that ever went forth to war' in 1914.[2]

Those who have retold this 'conventional story',[3] somewhat maligned as a 'triumphalist narrative',[4] have taken account of several caveats raised by contemporaries and by subsequent historians. These caveats may serve as a conceptual matrix since they almost certainly apply *mutatis mutandis* to the legacy of other colonial conflicts. First, the Boer War was described as anomalous and so few lessons for future wars could be drawn from it. The well armed, highly mobile Boers were not regular soldiers. After the loss of their capitals, they resorted to guerrilla warfare, trading space for time and using local conditions – atmospheric clarity, lack of natural obstacles and the expanse of the veld – in their long-range firing and 'hit and run' tactics. British mobility suffered, losing 350,000 horses (about 67 per cent of those employed) during the war as heavily-laden, English-bred horses disliked the 'thin, reedy, bitter grass' on the veld and succumbed to the deadly horse sickness of the wet months at lower altitudes.[5] Cavalry commanders, keen to exonerate their arm from wartime criticism, emphasised the abnormality arguments before the royal commissioners on the war, but, in making this case, Major-General J.P. Brabazon somewhat marred the argument by advocating that the main lesson of the war was the need to restore the war axe.[6]

Second, some emphasised the dangers of over-reaction as less than 400 men were killed in 'Black Week' compared with nearly one thousand at Isandlwana (1879) and a similar number at Maiwand (1880). Was not British expeditionary strategy based on 'strength in depth' and did not the army recover from these defeats as it had done on previous occasions and win a war over 6,000 miles from home?[7] Surely the war had positive aspects; Viscount Wolseley, the Commander-in-Chief, praised the call-out of reservists and the expeditionary achievement of sending 47,000 men to South Africa: these were 'the very ablest soldiers & thoroughly equipped for war'.[8] By contrast France, as Jay Stone notes, had to rent shipping from Britain in order to invade Madagascar in 1895, the Germans could not even send a battalion to the international operation on Crete in 1897, and the United States experienced severe logistical difficulties in invading Cuba in 1898.[9]

Third, Edwardian army reform was generated not simply on account of the South African experience. Ian Beckett, Stephen Badsey, Howard Bailes and others argue quite reasonably that the army was in a transitional phase even before the South African War. As drill books were being modified, tactical innovations practised and new forms of armament considered, the colonial experience, far from stimulating reform in some areas, merely hastened processes already in the pipeline.[10] Fourth, irrespective of whether reforms were generated by the war or simply accelerated by it, Tim Travers asserts that some of the worst features of the Victorian army persisted into the next conflict. He claims that the army 'learnt slowly' during the period 1900–14, that it altered some of the lessons of the Boer War and the Russo-Japanese War to fit a 'traditional ideal' and so suffered from a personalised command, with its damaging rivalries and frequent dismissals, and poor staff work leading to cover-ups in the Great War.[11] Beckett adds that the military interference in policy-making during the Boer War, examined so fully by Keith Surridge, was 'repeated on a far greater scale during the First World War'.[12] Finally, Haldane's claims that the army had been reformed from 1906 onwards with a continental commitment in mind have now been fully debunked, with widespread recognition that Britain's imperial commitments remained a primary concern for most of the pre-war period.[13]

All these caveats contain a kernel of truth, and in some cases much more than a kernel, but they hardly detract from the importance of the Boer War as catalyst for reform. The South African War, like all wars, was anomalous in some respects but it provided a major tactical challenge that would certainly dominate military concerns in the Great War, namely the crossing of fire zones swept by smokeless, flat-trajectory fire from magazine rifles. Although British forces had faced modern breechloaders on the Northwest

Frontier in 1897, the scale and shock of the South African experience was unprecedented. As Sir Neville Lyttelton remarked,

> Few people have seen two battles in succession in such startling contrast as Omdurman and Colenso. In the first, 50,000 fanatics streamed across the open regardless of cover to certain death, while at Colenso I never saw a Boer all day till the battle was over and it was our men who were the victims.[14]

If this was no ordinary war for the British army, the shock effects were magnified, as Stephen Badsey explains, because the Boer War was also a major media war with over seventy reporters at the front by early 1900: 'The Boer victories of "Black Week" ', he argues, 'derived their importance from their impact on politics and public opinion in London'.[15] This is certainly true but the repercussions were magnified partly because ministerial spokesmen shared Wolseley's confidence at the outset of the war, partly because the conflict became much more protracted than anyone had imagined, and partly because the counter-insurgency techniques – including farm-burning and concentration camps – became highly controversial and were denounced as 'methods of barbarism' by Sir Henry Campbell-Bannerman, the leader of the Liberal opposition.[16] This was not merely a colonial conflict but also a highly political and contentious conflict, one that established the issue of army reform, if not its precise scope and meaning, on the political agenda.

After the early defeats and the upsurge of political controversy, reform flourished both in South Africa and at home. The colonial context, though, hardly determined all the proposals, still less their evolution in the post-war years. British gunners, for example, had entered the war convinced that they needed European-style, quick-firing artillery. Wartime criticism that they had been out-ranged by the Boers, and humiliated at Colenso (where they lost ten guns in a precipitate forward deployment within 900 metres of the enemy without infantry support) was merely used by the Director-General of the Ordnance, Sir Henry Brackenbury, to persuade the government of the case for rearmament. However, the Boers had never used quick-firing guns, and the only quick-firers in South Africa were the naval guns (the 4.7 inch and 12-pounder) that were too heavy for field use. Accordingly an order was placed for German Erhardt guns and, after trials with them, a three-year programme of rearmament was launched in January 1901. In other words, the war was merely a means of justifying rearmament with a 13-pounder for horse artillery and the 18-pounder and 4.5 inch howitzer for field artillery. The perceived lessons from South Africa were very general, namely 'greater mobility for horse artillery, increased fire power for field artillery, and a longer range capability for both'.[17]

Similarly when Lord Roberts superseded Sir Redvers Buller in South Africa and issued his 'Notes for Guidance in South African Warfare' (26

January 1900), these tactical precepts, including careful reconnaissance before an attack, more use of cover and extended formations, avoiding artillery positions within range of an enemy's infantry, the use of continuous rather than sporadic bombardments, more marching and better care of horses by cavalry, and the delegation of responsibility to battalion and company commanders in the field,[18] bore all the hallmarks of lengthy service on the Northwest Frontier. Nor were these precepts uniquely understood by the Indian Army as the threat posed by smokeless, magazine rifles had been extensively debated in British military circles throughout the 1890s, and some officers had advocated changes in drill, training and tactics. The faulty tactics in South Africa, argues Howard Bailes, 'were not a consequence of the Aldershot teaching of the 1890s. They arose from a failure to act in accordance with it.'[19]

Like Badsey, who makes a similar point about some cavalry units practising dismounted action in the 1890s, they both agree that the Boer War erupted before these ideas became widely accepted.[20] This qualification is crucial: as the new tactical precepts were contentious, they were not part of a regularly practised doctrine. The home army had not engaged in large-scale manoeuvres for twenty-six years until the purchase of 41,000 acres on Salisbury Plain under the Manoeuvres Act of 1898. Hence while some cavalry units practised dismounted duties in the 1890s, the *Cavalry Drill* of 1898 allocated only five pages out of 460 to the topic and contrasted it with 'normal' mounted action. The new ideas had not convinced Lieutenant-Colonel Martin, the Commanding Officer of the 21st Lancers, who led his men on the disastrous charge at Omdurman and was quite unrepentant afterwards.[21] Nor had they persuaded Buller, the pre-war Adjutant-General who, after listening to a lecture on the possible effects on tactics of new weapons in February 1899, remarked 'when improvements are made in military arms and tactics they almost always follow along the same lines'.[22] Where the war provided a vital spur to reform was in discrediting some senior officers and in elevating others like Roberts and his supporters. During his eleven months in South Africa Roberts dismissed twenty-one senior officers, including eleven of the seventeen cavalry commanders, promoted reform before various royal commissions and select committees, and sought its implementation through post-war drill books and training.[23] He had political backing, too, as successive Secretaries of State – Brodrick, Arnold Forster and Haldane[24] – all endeavoured to meet public expectations, as expressed in Parliament and the press, about the need for army reform.

How effective were these reforms, bearing in mind that virtually every facet of the military system came within the review of various royal commissions and parliamentary committees? There were inquiries into issues such as the degeneration of the imperial race, the education and training of officers, the War Office, and the Militia and Volunteers. The royal commission on the war, chaired by the Earl of Elgin, sat for fifty-five days and asked

22,000 questions of 114 military and civilian witnesses. Its bulky, two-volume report, published in August 1903, may not have proposed many reforms but it provided plenty of ammunition for reformers. One commissioner, Lord Esher, a confidant of the king and of the Prime Minister, Arthur Balfour, produced a more forthright set of recommendations in subsequent reports from his War Office (Reconstitution) Committee (January and March 1904).

Once again the specific reforms of higher defence organisation and the War Office had little to do with the South African War. Both the concepts of a Committee of Imperial Defence and a General Staff had been raised by the Hartington Commission when it inquired into the 'Civil and Military Administration of the Naval and Military Departments' and reported in 1890. The war simply enabled resistance to these ideas to be swept aside: the Committee of Imperial Defence (CID) was formed by Balfour in 1902 (and later gained a permanent secretariat) and the post of Commander-in-Chief was abolished in 1904 to be replaced by a Chief of the General Staff (CGS). In due course the 'managerial revolution' in the War Office was completed by the creation of a General Staff and an Army Council.[25]

However, none of these reforms ensured that Britain would be better prepared for a continental conflict. In the first place, the imperial focus predominated, with the CID devoting fifty of its first eighty-two meetings to the defence of India. Even after the Anglo-Russian agreement of 1907 the CID continued to assume that 100,000 men might need to be sent to India, and the defence of India against a potential Russian threat remained a prime concern until 1910 at least.[26] Second, the institutions were only as good as the men appointed to serve in them, and the first CGS, Sir Neville Lyttelton, proved, as John Gooch observes, a 'disastrous' appointment.[27] His successors were more effective, and two Directors of Military Operations, John Spencer Ewart and Henry Wilson, were 'Continentalists by orientation'. Their efforts, particularly those of the francophile Wilson, greatly enhanced Britain's preparations for mobilisation in 1914, belying the charge that 'All Staff work in the decade before 1914 was bedevilled at every turn by the conflict between the new continental commitment and the continuing colonial one.'[28] But the General Staff, which existed only within the War Office until 1906, never inculcated a common doctrine that would facilitate the employment of large formations in the field, the large formations that would be essential for any war in Europe.

Third, the reform programme suffered from peacetime political realities, namely financial constraints and opposition to any form of compulsory service in peacetime. Haldane insisted that all his army reforms had to fit within a tight financial ceiling of £28 million and regarded his Territorial Force as an alternative to the compulsory service being advocated by the National Service League.[29] Accordingly, the limitations that had always bedevilled peacetime training in a small voluntary army recurred, especially as the army's strength fell from 212,393 on 1 January 1898 to 192,144 by

1913. Men were in short supply as soldiers performed non-military duties in barracks as cooks, waiters, servants, and labourers (7,000 on permanent employments and 10,000 on daily casual employments in 1902).[30] Under the linked-battalion system, in which half of the army served in India or other overseas garrisons, home battalions provided drafts annually for their linked battalions overseas. Under the short-service system of recruiting, they also had to instruct large numbers of recruits each year that were unavailable for field training. Officers never commanded full-strength units in peacetime, and sometimes trained two units in one or occasionally considered theoretical schemes with imaginary troops. Training, as Sir William Robertson remarked, was 'largely a case of trying to make bricks without straw'.[31] Only in the manoeuvres of 1910 was one of the two Aldershot divisions mobilised at full strength, but then only at the expense of the other division and by securing volunteers from the First Class Army Reserve. In the Army Exercise of 1913, Sir John French theoretically commanded four divisions and a cavalry division, but a mere 47,000 men participated, with the battalions averaging 300 men apiece (less than half of their peacetime establishments) and with field batteries at half-strength. Manpower shortages, in sum, devalued unit training at all levels, undermined the benefit of divisional training, and never permitted any practical training at corps level.[32]

Finally, there was the pervasive refrain that Britain should learn 'the lessons of the war' and adopt wholesale reform. Travers doubts that this process succeeded, claiming that the Edwardian army revived an 'amateur, traditional ideal of war', with an emphasis upon moral qualities and the imperative of mounting offensives, but he is perhaps too severe.[33] Quite apart from the difficulty of defining terms such as 'amateur' or 'traditional', there were not any self-evident lessons about major wars to be deduced from the South African conflict. Clear unambiguous lessons could not be easily deduced from tactical encounters in colonial circumstances. What did the bombardment of hastily erected trenches at Paardeberg prove? Were long-range bombardments basically ineffective since ninety-eight guns pounded these positions over eight days but caused only 100 casualties, or was this further confirmation that British guns and ordnance were obsolete, and that Britain needed to rearm with high-angled howitzers? Similarly, the ability of the Boers to disengage and withdraw without pursuit at Zand River (May 1900) prompted Sir Ian Hamilton to describe it as a 'fiasco', a humiliation, and a culminating incident that eroded his faith in the efficacy of shock tactics. Conversely Sir John French, a redoubtable cavalry commander, hailed the encounter as a splendid triumph for the moral force of cold steel.[34]

The advent of long-range, smokeless, flat-trajectory fire from magazine rifles had a radical effect upon the offence and defence. While it confounded frontal attacks in close order, the forward deployment of artillery pieces, and massed knee-to-knee cavalry charges (and the cavalry never mounted such charges in South Africa), it offered new opportunities for defensive fighting.

Some trench lines were moved off hilltops into valleys, and camouflage, smokeless powder, longer-range or rapid firepower increased the difficulty of scouting properly and estimating enemy numbers. Night operations facilitated both offensives and withdrawals, and the larger, more deadly fire zones prompted recourse to wider extensions and deeper deployments (at Poplar Grove, each brigade of two battalions formed a double column of extended companies, with a total frontage of about half a mile and a depth of over a mile).[35] In these larger and deeper battlefields, where advances were increasingly made under the cover of rolling barrages (first demonstrated at Pieter's Hill), attacks required the delegation of command to lower echelon officers and non-commissioned officers, with camouflage-clad infantry becoming more adept at finding cover, advancing by rushes, firing rapidly, and digging-in when necessary.

In facilitating these diverse and evolving operations over a vast territorial area, the army's support services, apart from the Royal Army Medical Corps, performed remarkably well. All services were understaffed and under-resourced for a war of this magnitude (with the Royal Engineers originally deploying 694 officers and men in South Africa on 1 October 1899 only to find that it needed 7,000 officers and men during the war as a whole and another 1,500 engineers from the Militia and Volunteers). Most of these services had to employ untrained men from other units or large numbers of civilians, but the RAMC found itself hopelessly 'overworked, undermanned and under-orderlied'.[36] It failed to cope effectively when a typhoid epidemic swept through the camp at Bloemfontein and became increasingly dependent upon civilian assistance. Expanding these services was one of the least ambiguous lessons of the war, but the army also learned useful lessons in military railway management, medical practice, ambulance design, and hospital trains. There were also minor improvements in battle dress, including the use of the puttee, khaki-coloured clothing with aprons for kilts, aluminium canteens, new water carts and looped cartridge belts.[37]

Offensives increasingly involved attempts to pin the enemy frontally, while mounting flank attacks (for which accurate reconnaissance was essential but difficult to supply) and then, if the Boers withdrew (as all too often happened), a capacity to pursue were both roles for which a much heavier cavalry, armed with lance or sword and carbine, was ill equipped. The cavalry in spite of some notable successes (harrying retreating Boers at Elandslaagte, mounting an open-order charge against a weakly held position at Klip Drift, and later trapping and holding Cronje's Boers at Paardeberg) struggled initially in its scouting and pursuit missions (notably at Modder River and Poplar Grove). Although it improved its scouting techniques, and enhanced its firepower by replacing the carbine with the rifle, the army came to rely increasingly upon mounted infantry in the second phase of the war when it mounted a series of drives against a highly mobile adversary.[38]

Yet the cavalry, unlike the other two arms, never suffered a catastrophic reverse, and as cavalry engagements were conspicuous by their absence in

Manchuria, there were not any self-evident lessons for the future. So commentators interpreted the colonial experience to suit their subjective preferences. When Roberts was Commander-in-Chief, he armed the cavalry with the same magazine rifle as issued to the infantry – the Lee Enfield, shortened and lightened to suit cavalry requirements. He withdrew the lance as an active service weapon and, in a signed preface to *Cavalry Training, 1904*, insisted 'that instead of the firearm being an adjunct to the sword, the sword must henceforth be an adjunct to the rifle'.[39] When Roberts and his supporters fell from office, Douglas Haig, as Director of Military Training then Director of Staff Duties (1906–9), and French, as Inspector-General of the Forces (1907–11), reversed these priorities. Neither opposed dismounted action in principle; both had advocated it in the early 1890s and accepted that the arm had to become proficient in mounted and dismounted duties, but they insisted that the Boer War was abnormal. They argued that the terrain in Europe could facilitate surprise attacks, that they had to confront cavalries equipped for the *arme blanche*, and that the essence of cavalry spirit could only be based on the lance and sword. As Haig remarked on the cavalry staff rides of 1909, the effects of a 'skilfully managed dismounted action' would never compare with the rare though decisive charges with the *arme blanche*.[40]

The artillery faced similar problems. Whereas rearmament with quick-firing guns enjoyed widespread support, there was less agreement about the tactical implications of these guns, still less about the lessons of the South African experience. Reformers pressed for longer-range bombardments, more use of cover and dispersed batteries, and tactics that aimed, in support of an infantry advance, not to clear trenches but to confine defenders and distract them from advancing infantry.[41] Even these relatively modest reforms incurred opposition from officers who regarded the topographical and climatic conditions in South Africa as abnormal (and hence had doubts about long-range bombardments), and who feared that the quest for cover and indirect fire would impair both the accuracy of artillery fire and the offensive zeal of the gunner. Yet the reformers only selected certain lessons from South Africa, and, in the first post-war manual, retained traditional concepts such as the artillery duel and preliminary bombardments (despite Magersfontein where a preliminary bombardment, averaging 1,047 rounds per battery, merely wounded three Boers and left their entrenchments undisturbed).[42]

Artillery reform would evolve thereafter, paying some heed to the Manchurian experience but rather more to French tactical developments, government parsimony and pressure from the other arms. The Far Eastern spectacle of organised and trained artilleries exploiting covered positions and smokeless powder, albeit without any quick-firing ordnance, was extensively reviewed in Britain. If Japanese gunners earned plaudits for their mobility, ability to fire from covered positions, and readiness to support infantry assaults by night and day until enemy trenches were taken, the consumption of ammunition on both sides aroused acute concern. As

Colonel F.D.V. Wing said of the Russian use of 120,000 rounds, weighing about 800 tons, in four days at Liaoyang, 'the power of expenditure far exceeds the power of supply'.[43] This particularly applied in the United Kingdom where the government was unwilling to expand the production of artillery shells when it was already committed to the Dreadnought-building programme. Accordingly when British observers lauded the rapid firing of French four-gun batteries at medium ranges (exploiting an automatic fuze setter and a more stable carriage than the British QF guns), Sir Ian Hamilton quickly reproved the military correspondent of *The Times* lest he prompt the government to accept the four-gun standard and

> refuse to create any more new four gun batteries, and thereby reduce expense and guns by one third. So whatever you do, my dear Repington, for God's sake keep quiet about this idea of four gun batteries.[44]

What British gunners accepted from Japanese practice, French theory, and pressure from other arms was that the classical artillery duel and independent artillery action were now impractical under modern conditions. They appreciated, too, that their main objective had to be the provision of tactically useful fire in support of the infantry offensive. The doctrine of 'Economy of Force', embodied in the manuals of 1912 and 1914, recognised that batteries could be held in observation, or even in reserve until needed; that more rapid fire, and not necessarily a greater number of guns, could increase the volume of fire; and that the infantry and artillery could operate in tactical groups, linked by an advance observation officer. The manuals explicitly stated that the artillery should not be inhibited by the danger of shells falling short in their shelling of the enemy during an infantry attack.[45]

All arms, not least the infantry, were hampered in their reform by the constraints imposed upon peacetime funding. For example, Lord Roberts sought to improve British firepower by increasing the ammunition allotted to annual musketry practice, by diminishing collective practice and volley-firing, and by increasing individual firing at fixed and mobile targets over short to medium ranges. During the training year 1902–3, the Conservative government doubled the allocation to cavalry and increased the infantry allocation by a third but, by 1907, when the surplus stocks of wartime ammunition were exhausted, the Liberal government duly rescinded the additional allowance. Whether the School of Musketry's case for the allocation of additional machine guns per battalion was thwarted on financial grounds as Sir James Edmonds argues, or was rebuffed by the General Staff directors lest it inhibit movement on the offensive, as Bidwell and Graham claim, the effect was the same: the army had to compensate by improving individual rates of rapid fire. By 1911 British soldiers achieved a best rate of fifteen rounds a minute, while the Japanese regarded eight rounds a minute as excessive and the conscript soldiers of France, Germany and Russia only attained twelve rounds a minute at under 300 yards.[46]

The essence of post-war tactics was preserving the ability to manoeuvre in the face of enhanced firepower. In South Africa the infantry had dramatically improved their fieldwork, use of ground and shooting at various ranges. Practice and experience had enabled soldiers to become highly proficient, prompting reformers to argue that the post-war imperatives included some means of sustaining the improvement in musketry, preferably with an improved rifle, and more practice in taking cover, digging entrenchments and moving in extended formations.[47] There were differing views about how best to do this in peacetime training, and some doubts about the ability of the rank-and-file, in view of their educational limitations and predominantly urban origins, to adapt to the demands of modern warfare.[48] The War Office was also aware that the government, having just bought additional ground on Salisbury Plain, was unlikely to purchase further tracts of land for target practice, still less to provide the substantial increase in army pay and allowances that might broaden the appeal of service life and attract a better quality of recruit. In revising peacetime training it had to operate within financial constraints and utilise existing facilities.

It succeeded to a remarkable degree. Training became a cumulative process, beginning with individual instruction in the winter; followed by squadron, company, and battery training in the spring; regimental, battalion and brigade training in the summer; and, finally, divisional or inter-divisional exercises and army manoeuvres in late summer and early autumn. Soldiers became increasingly proficient in the use of the new Short Lee Enfield rifle and adept at both fire and movement in the attack and in regulated fire from defensive positions. These exercises were practised recurrently by squads or sections attacking each other in company training, often without orders from officers and non-commissioned officers so that the ranks would act increasingly on their own initiative.[49] At the Army Exercise of 1913 Commandant de Thomasson observed that the infantry 'makes wonderful use of the ground, advances as a rule by short rushes and always at the double, and almost invariably fires from a lying down position'.[50]

The priority, as hammered home in manuals, pronouncements by the General Staff and speeches by senior officers, was to assume the offensive. Fire action, though vitally important, was seen as subordinate to forward movement. The firing line was supposed initially to advance as far as possible without opening fire to conceal its approach and conserve ammunition. Thereafter it had to attack in successive waves, regarding fire positions as transitory objectives, with the aim of closing upon the enemy lines. The doctrine sought to maximise the assailant's moral advantages – his initiative, freedom of action and power of manoeuvre – and so overcome the fire effect of contemporary weapons from defensive positions.[51] Such thinking was not simply a reversion to 'traditional' principles nor an attempt to compensate for the perceived moral shortcomings of the Edwardian ranker,[52] but reflected the spurious wisdom gleaned from Paardeberg and the Japanese victories at Liaoyang and Mukden, and an understanding of modern war that cohered with the tactical assumptions of the French and German armies. The question was not whether but how to attack, and the General Staff eschewed any rigid adherence to the tactics of penetration (preferred by the French) or the tactics of envelopment (preferred by the Germans). The British army chose pragmatism as it could only deploy six divisions and one cavalry division, a tiny force by comparison with its continental neighbours. The *Memorandum on Army Training 1910* declared that

> success depends not so much on the inherent soundness of a principle or plan of operations as on the method of application of the principle and the resolution with which the plan is carried out.[53]

In some areas, namely officer education, reforms were relatively modest. Where the army controlled the process, as in the curriculum of the Staff College, sweeping changes in content and assessment could be completed under another of Roberts's protégés, Henry Rawlinson (commandant, 1903–6). But for the regimental officer little could be done without the support of the public schools, and the latter would brook no deviation from the principle of nonspecialisation in public service examinations.[54] In providing a means of expanding the British Expeditionary Force (BEF), Haldane was more ambitious. He sought to create a special reserve from the Militia, a reserve of officers from the new Officers' Training Corps, and a Territorial Force from the Volunteers. If political compromises riddled these reforms (the Territorials were formally raised for home defence), and none of them met their full establishments, they represented an organised framework of expansion, reasonably complete in arms, equipment and support services, albeit a framework discarded by Kitchener, who knew little of the home army, when he became Secretary of State for War in August 1914.

Nevertheless, the BEF had certainly benefited immeasurably from the recent combat experience, an experience that fed its way into post-war training at individual, collective and inter-arm levels. That training may have lacked a specific continental focus and, under the constraints of voluntary service, may not have been able to function above divisional level, but the standards achieved were remarkable. As Henry Wilson commented after observing the French manoeuvres of 1913, 'The French Inf[antry] are marvels of endurance & good spirits but we have nothing to learn from them in handling the actual troops on the ground'.[55] Where the South African experience was less helpful was in planning for the carnage of large-scale warfare. British wastage calculations, set at 40 per cent for the first six months and between 65 and 75 per cent of the first twelve months, reflected an average based upon the South African and Manchurian experiences. The actual casualty rate would prove to be 63 per cent within the first three months, prompting Cyril Fall to comment on the BEF: '*Armées d' élite* would be invincible if wars were fought without casualties. Things being what they are, *armées d' élite* are unlikely to remain so for long.'[56]

Acknowledgements

I should like to thank The Trustees of the Liddell Hart Centre for Military Archives, the Imperial War Museum, National Army Museum, Brighton and Hove City Council, and the National Library of Scotland for permission to quote from papers which are held in their archives.

Notes

1 Rudyard Kipling, 'The Lesson', *The Times*, 29 July 1901, p. 6.
2 Sir James Edmonds, *Official History of the Great War: Military Operations, France and Belgium, 1914*, hereafter referred to as *OH* , London: HMSO, 1922, 1, pp. 10–11.
3 Edward M. Spiers, 'The Late Victorian Army, 1868–1914', in David Chandler and Ian Beckett (eds) *The Oxford Illustrated History of the British Army*, Oxford: Oxford University Press, pp. 189–214; Ian F.W. Beckett, 'The South African War and the Late Victorian Army', in Peter Dennis and Jeffrey Guy (eds) *The Boer War: Army, Nation and Empire*, Canberra: Army History Unit, 2000, pp. 31–44.
4 Hew Strachan, 'The Boer War and Its Impact on the British Army, 1902–14', in Peter B. Boyden, Alan J. Guy and Marion Harding (eds) *Ashes and Blood: The British Army in South Africa 1795–1914*, London: National Army Museum, 1999, pp. 85–98.
5 Stephen Badsey, 'Mounted Combat in the Second Boer War', *Sandhurst Journal of Military Studies*, 2 (1991), pp. 11–27; Ian Beckett, 'Military High Command in South Africa, 1854–1914' in Boyden, Guy and Harding (eds) *Ashes and Blood*, pp. 60–71.
6 PP, *Minutes of Evidence Taken before the Royal Commission on the War in South Africa*, hereafter referred to as the *Elgin Commission*, Cd. 1790 (1904), XL, Major-General J.P. Brabazon (Qs. 6,841–2 and 6,915–20); see also Cd. 1791 (1904), XLI, Sir John French (Q. 17,240).
7 Stephen Badsey, 'The Boer War as a Media War', in Dennis and Guy (eds) *The Boer War: Army, Nation and Empire*, pp. 70–83.
8 Royal Pavilion Libraries and Museums, Brighton and Hove City Council, Hove Library, Wolseley Mss., W/P 28/61, Viscount Wolseley to Lady Wolseley, 29 September 1899.
9 Jay Stone, 'The Anglo-Boer War and Military Reforms in the United Kingdom', in Jay Stone and Erwin Schmidl, *The Boer War and Military Reforms*, Lanham MD: University Press of America, 1988, p. 25.
10 Beckett, 'South African War and the Late Victorian Army', p. 44; Badsey, 'Mounted Combat', p. 14; Howard Bailes, 'Technology and Tactics in the British Army, 1866–1900', in Ronald Haycock and Keith Neilson (eds) *Men, Machines and War*, Waterloo, Ontario: Wilfred Laurier University Press, 1988, pp. 23–47; Edward M. Spiers, 'Rearming the Edwardian Artillery', *Journal of the Society for Army Historical Research*, LVII: 231 (1979), pp. 167–76.
11 Tim Travers, *The Killing Ground: The British Army, the Western Front and the Emergence of Modern Warfare, 1900–18*, London: Allen & Unwin, 1987, pp. 27, 36n. 105; see also Philip Towle, 'The Russo-Japanese War and British Military Thought', *Journal of the Royal United Services Institute for Defence Studies*, 116 (1971), pp. 64–8; Keith Neilson, "That Dangerous and Difficult Enterprise": British Military Thinking and the Russo-Japanese War', *War and Society*, 9: 2 (1991), pp. 17–37.
12 Beckett, 'South African War and the Late Victorian Army', p. 44; Keith T. Surridge, *Managing the South African War, 1899–1902: Politicians versus*

Generals, Woodbridge: Boydell Press for the Royal Historical Society, 1998, pp. 175–84; and 'Rebellion, Martial Law and British Civil-Military Relations: The War in Cape Colony, 1899–1902', *Small Wars and Insurgencies*, 8: 2 (1997), pp. 35–60; see also Denis Judd and Keith Surridge, *The Boer War*, London: John Murray, 2002, chs 13, 21–4.

13 Edward M. Spiers, *Haldane: An Army Reformer*, Edinburgh: Edinburgh University Press, 1980, pp. 11–28; Keith Neilson, *Britain and the Last Tsar: British Policy and Russia, 1894–1917*, Oxford: Clarendon Press, 1995, pp. 126–36; John Gooch, *Plans of War: The General Staff and British Military Strategy c. 1900–1916*, London: Routledge and Kegan Paul, 1974, pp. 198–232; Strachan, 'Boer War and Its Impact', pp. 87–90.

14 General Sir Neville Lyttelton, *Eighty Years: Soldiering, Politics, Games*, London: Hodder and Stoughton, 1927, p. 212.

15 Badsey, 'Boer War as a Media War', pp. 79, 81; see also W.S. Hamer, *The British Army: A Study in Civil-Military Relations*, Oxford: Clarendon Press, 1970, pp. 174–5.

16 Edward M. Spiers, *The Late Victorian Army, 1868–1902*, Manchester: Manchester University Press, 1992, pp. 310–11; *The Times*, 15 June 1901, p. 12.

17 Spiers, 'Rearming the Edwardian Artillery', p. 171.

18 National Army Museum (NAM), 7101–23–111–1, Lord Roberts, 'Circular Memorandum No. 5, Notes for Guidance in South African Warfare', 26 January 1900.

19 Bailes, 'Technology and Tactics', pp. 39, 46 and 'Technology and Imperialism: A Case Study of the Victorian Army in Africa', *Victorian Studies*, 24 (1980), pp. 82–104.

20 Bailes, 'Technology and Tactics', p. 46; Badsey, 'Mounted Combat', p. 14.

21 *Cavalry Drill*, London: HMSO, 1898, pp. 386–91;'The Charge of the Lancers at Omdurman', *Hampshire Chronicle*, 15 October 1898, p. 6; see also Edward M. Spiers, 'Campaigning under Kitchener', in Edward M. Spiers (ed.) *Sudan: The Reconquest Reappraised*, London: Croom Helm, 1998, p. 72.

22 Sir Redvers Buller's comments on Brevet Lt.-Col. F.B. Elmslie, 'The Possible Effect on Tactics of Recent Improvements in Weapons', *Aldershot Military Society*, 72 (6 February 1899), p. 18.

23 NAM, Acc. 7101–23–122–2, Roberts to A. Akers-Douglas, 29 August 1901; evidence before the *Elgin Commission* (qs. 10,333–35, 10,409, 10,441–47).

24 Lowell J. Satre, 'St John Brodrick and Army Reform, 1901–1903', *Journal of British Studies*, 15 (1976), pp. 117–39; Albert V. Tucker, 'The Issue of Army Reform in the Unionist Government, 1930–5', *Historical Journal*, 9 (1966), 90–100; John Gooch, 'Haldane and the "National Army"', in Ian F.W. Beckett and John Gooch (eds) *Politicians and Defence: Studies in the Formulation of British Defence Policy*, Manchester: Manchester University Press, 1981, 69–86.

25 John Gooch, 'Britain and the Boer War' in George J. Andreopoulos and Harold E. Selesky (eds) *The Aftermath of Defeat: Societies, Armed Forces, and the Challenge of Recovery*, New Haven CT: Yale University Press, 1994, pp. 40–58.

26 Gooch, *Plans of War*, pp. 198–232.

27 Gooch, 'Britain and the Boer War', p. 55; Shelford Bidwell and Dominick Graham, *Fire Power British Army Weapons and Theories of War 1904–1945* London: George Allen & Unwin, 1982, p. 41.

28 Ibid. and Strachan, 'Boer War and Its Impact', pp. 89–91.

29 Spiers, *Haldane*, pp. 48–73, 168–83, 198–9.

30 The National Archives, Public Record Office, W.O. 32/9120, Adjutant-General, 'Minute on replacement of soldiers on non-military duties by ex-soldiers or civilians', 1 October 1902.

31 Field Marshal Sir William Robertson, *From Private to Field Marshal*, London: Constable, 1921, p. 159.

32 Edward M. Spiers, 'Reforming the Infantry of the Line, 1900–1914', *Journal of the Society for Army Historical Research*, LIX: 238 (1981), pp. 82–94; Strachan, 'Boer War and Its Impact', p. 90.

33 T.H.E. Travers, 'Technology, Tactics and Morale: Jean de Bloch, the Boer War, and British Military Theory, 1900–1914', *Journal of Modern History*, 51 (1979), pp. 264–86; and 'The Hidden Army: Structural Problems in the British Officer Corps, 1900–1918', *Journal of Contemporary History*, 17 (July 1982), pp. 523–44. But officers ignored some lessons from South Africa by restoring conspicuous badges of rank, swords and impractical hats for service in 1914, K. Simpson, 'The officers' in I.F.W. Beckett and K. Simpson (eds) *A Nation in Arms* Manchester: Manchester University Press, 1985, p. 84.

34 Liddell Hart Centre for Military Archives, King's College London, Hamilton Mss., 5/1/8, Sir Ian Hamilton to Erskine Childers, 30 October 1910; Sir John French, 'Preface', in General F. von Bernhardi, *Cavalry in War and Peace* London: Hugh Rees, 1910, p. xi.

35 Maj-Gen. Sir Frederick Maurice and Captain M.H. Grant, *History of the War in South Africa*, 4 vols, London: Hurst & Blackett, 1906–10, 2, p. 198.

36 Sir H.M. Leslie Rundle, q. 17,924 evidence before the *Elgin Commission*.

37 Spiers, *The Late Victorian Army*, pp. 322–5; Andrew H. Page, 'The Supply Services of the British Army in the South African War, 1899–1902', unpublished D.Phil. thesis, Oxford, 1976, pp. 336–47; Stone, 'Anglo-Boer War and Military Reforms', pp. 57–69.

38 Stone, 'Anglo-Boer War and Military Reforms', pp. 36–7, 80–3, 92–4.

39 Lord Roberts, 'Preface', in *Cavalry Training, 1904*, London: HMSO, p. v; see also Brian Bond, 'Doctrine and Training in the British Cavalry, 1870–1914', in Michael Howard (ed.) *The Theory and Practice of War: Essays Presented to Captain B.H. Liddell Hart*, London: Cassell, 1965, pp. 95–125; Edward M. Spiers, 'The British Cavalry, 1902–14', *Journal of the Society for Army Historical Research*, LVII (1977), pp. 71–9.

40 National Library of Scotland, Acc. 3155, Haig Mss., Vol. 82, Sir Douglas Haig, 'Reports on Cavalry Staff Rides 1909 held 1–6 March 1909'; Badsey, 'Mounted Combat', pp. 11–28.

41 NAM, Acc. 7101–23–124–3, Roberts Mss., Roberts, 'Minute on Tactical Training of Horse and Field Artillery', 21 November 1902; *Field Artillery Training 1902* London: HMSO, 1902, ch. 1.

42 Ibid.; Sir George Marshall, q. 18,517 evidence before the *Elgin Commission*; Spiers, 'Rearming the Edwardian Artillery', pp. 173–4.

43 Colonel F.D.V. Wing, 'The Distribution and Supply of Ammunition on the Battle-Field', *Journal of the Royal United Service Institution*, 52 (1908), pp. 895–924; Major C.E.D. Budworth, 'Tactical Employment of Artillery as Evolved on the Practice Ground and from the Experiences of Modern War', *Aldershot Military Society*, XCIX (16 February 1908), pp. 8–10.

44 Liddell Hart Centre for Military Archives, King's College London, Hamilton Mss., 5/1/7, Hamilton to Repington, 27 October 1910.

45 *Field Artillery Training 1914*, London: HMSO, 1914, ch. 7.

46 Lieutenant-Colonel, J. Campbell, 'Fire Action', *Aldershot Military Society*, CXII (14 March 1911), pp. 5–6; see also Captain R.V.K. Applin, 'Machine Gun Tactics in our Own and Other Armies', *Journal of the Royal United Services Institution*, 52 (1908), pp. 34–65; *OH*, 2, p. 463; Bidwell and Graham, *Fire Power*, pp. 49–54.

47 Roberts, q. 10,442 and Sir Ian Hamilton, q. 13,941 evidence before the *Elgin Commission*; Spiers, 'Reforming the Infantry', pp. 85–6.

48 Sir Thomas Kelly Kenny, q. 16,924, evidence before the *Elgin Commission*.

49 Campbell, 'Fire Action', p. 7.
50 Commandant de Thomasson, 'The British Army Exercise of 1913', *Army Review*, 6 (1914), pp. 143–56.
51 Campbell, 'Fire Action', pp. 1 and 5; Lieutenant-Colonel N.R. McMahon, 'Fire Fighting', *Aldershot Military Society*, XCV (18 February 1907), pp. 4–11, 18.
52 Travers, 'Offensive and the Problem of Innovation', pp. 538–9, 546–7.
53 'The British Army and Modern Conceptions of War', *Edinburgh Review*, 113 (April 1911), pp. 321–46; Spiers, 'Reforming the Infantry', pp. 91–2; Neilson, 'That Dangerous and Difficult Enterprise', p. 20.
54 Gooch, 'Britain and the Boer War', pp. 55–6.
55 Imperial War Museum, Wilson Mss., Major-General Henry H. Wilson, diary, 14 September 1913.
56 Cyril Falls, *The First World War*, London: Longmans, 1960, p. 16; Edward M. Spiers, *The Army and Society 1815–1914*, London: Longman, 1980, pp. 293–5.

2 Big wars and small wars between the wars, 1919–39

David French

By the 'Hundred Days' campaign in 1918 the BEF had developed a sophisticated combined arms doctrine and practice that enabled it to inflict a
crushing defeat on the German army. The inter-war General Staff seemed to
take this lesson to heart. Speaking after an exercise in 1927, the Chief of the
Imperial General Staff (hereafter CIGS), Sir George Milne, insisted that 'It
is the co-operation of all necessary arms that wins battles and that is your
basis for training for the future. I want that to be your principle in training –
combination and co-operation of arms.'[1] But the experience of the first half
of the Second World War seemed to demonstrate that at some point in the
inter-war period this precept had been forgotten. There were many reasons
for the often lacklustre performance of the British army between 1939 and
1942. Not the least of them was the helter-skelter expansion that it underwent between 1939 and 1941 that led to a dangerous dilution of its trained
pre-war cadre. But what must not be overlooked was that between the wars the
cadre itself had been inadequately trained to practise combined arms tactics.

The focus of this book is the extent to which the army's commitment to
fighting small wars and colonial insurgencies prevented it from preparing
adequately for 'big wars'. The focus of this chapter is the extent to which the
everyday requirements of colonial soldiering in the 1920s and 1930s
prevented the army from preparing for the kind of war it found itself
fighting after 1939. Between the wars the General Staff was acutely aware
that this was a potential problem. As the 1930 edition of the *Field Service
Regulations (Organization and Administration)* (hereafter *FSR*) explained,

> the British Empire is confronted with problems peculiarly its own.
> Unlike a continental power, it consists mainly of a number of self-
> governing communities widely separated and of varying resources. In
> defence of its vital interests it may be called upon to place a force in the
> field under conditions varying from a small expedition against an
> uncivilised enemy to a world-wide war.[2]

The General Staff's solution was outwardly simple. 'The instructions laid
down herein', explained *FSR* (1929), 'cover a war of the first magnitude, but

are to be modified in their application to other forms of warfare.'[3] The meaning they ascribed to a 'war of the first magnitude' was obvious from the contents of the manual. Only one of its twelve chapters dealt with what the authors called 'Warfare in undeveloped and semi-civilised countries'. The remainder assumed that the British army would be pitted against forces at least as well equipped and trained as it was.

But the everyday reality between the wars was that the British army was far more likely to find itself committed on the ground to fighting small wars and colonial insurgencies. A cursory examination of its more significant operations suggests that troops took part in attempting to suppress at least five significant colonial insurgencies (Egypt, 1919; Ireland, 1919–21; Iraq, 1920; Palestine, 1929 and 1936–9), two 'small wars' (Afghanistan, 1919; Turkey, 1919–22) as well as being called upon to give 'aid to the civil power' in the UK, the Caribbean, Cyprus, China, the Sudan and in India on numerous occasions.[4]

The purpose of this chapter is, therefore, to estimate the extent to which preparing for and conducting these operations did impede the army's ability to prepare for and to fight the 'big' war that was to confront it after 1939. It will do so by looking in turn at the conceptual base, the material base and the personnel base, the three components of the army that came together to produce its combat capability.[5]

The conceptual base

The General Staff's *Training Regulations, 1934* echoed the insistence of *FSR* that the British army had to prepare for a multiplicity of different missions. It listed four broad categories of operations that troops might find themselves engaged upon: 'Imperial policing'; 'minor expeditions, possibly on peace establishments'; 'Major expeditions which may or may not include the Territorial Army'; and finally 'a national war'.[6] At any one time the Cardwell system meant that about half of the army was stationed in overseas garrisons where local commanders naturally gave priority to training for the most likely missions that would confront their own units. The dilemma facing the General Staff was which of these types of operations it ought to give priority to when it issued its training instructions to govern the activities of the other half of the army, those units which were retained in Britain, partly to feed their linked units overseas, and partly to form the cadre of the regular expeditionary force. These priorities changed over time, but, except for a brief period in the early 1930s, it never entirely lost sight of the need for troops to prepare for a 'national war'.

In the first half of the 1920s the heaviest emphasis was placed on the need to prepare for wars outside Europe. In February 1922 the General Staff told all Commands that they were to train 'for the most probable war, i.e. against an enemy rather worse armed than ourselves, with comparatively small forces, and using existing resources as laid down in F.S.R.'.[7] A few months later the new CIGS, the Earl of Cavan, reminded senior officers that they should train for

> mobile warfare; warfare, that is, of the type in which our army has in the past most frequently been engaged, as witness the campaigns in S[outh]. Africa, in Waziristan, against the Moplahs, or in Iraq during General Maude's final advance on Baghdad.[8]

However, that did not mean that Cavan wanted his subordinates to ignore entirely the possibility of fighting a European enemy. He recognised that peace in Europe might not endure, and that, although the army's next enemy might not be as well equipped as the British, it ought to assume that they would be. In a way that echoes contemporary doctrine, Cavan insisted that the army had to train on the assumption that their enemies would be as well equipped as they were because it was the only way of ensuring that the British army could keep abreast of modern developments.[9]

The need to prepare for the more immediate possibility of fighting on the frontiers of the empire, but not to lose sight of the more remote possibility that the army might have to engage in another European war, underpinned the General Staff's training instructions for the army in Britain until the early 1930s. Despite the Ten Year Rule and the Locarno Pact, much of the everyday training of the army at home and the various experimental organisations that Cavan's successor, Sir George Milne, established were directed towards preparing the army to fight another European power. In 1928 the Director of Military Training told a General Staff conference that troops in Britain 'train for a mobile war against an organised enemy armed with rifles, machine guns, guns, aeroplanes and tanks, and possibly gas'.[10] The need to do just that had gone so far that by 1932, according to one brigade commander, the army at home was so concerned 'in the chase after European war and the shibboleths of "Fire plans" etc', that they found the mental adjustments required to fight a second-class enemy outside Europe difficult to accomplish.[11]

It was only for a very brief period in the early 1930s, following the signing of the Kellog-Briand Pact, and in the wake of the drive for international disarmament that culminated in the convening of the Geneva Disarmament Conference in February 1932, that the General Staff ordered the regular army at home to concentrate its training exclusively on imperial operations.[12] *Army Training Memorandum Number 4A*, issued in December 1931, insisted that the army's first training priority was imperial policing, its second was minor expeditions, then major expeditions, and preparing for a great national war came last.[13] But by 1934 Milne's successor, Sir Archibald Montgomery-Massingberd, had reversed these priorities. Persuaded by Hitler's accession to power and Germany's withdrawal from the League of Nations and the Disarmament Conference of Nazi Germany's dangerous intent, Montgomery-Massingberd was determined to modernise the army to prepare it for a continental land war.[14] In September 1934 the Director of Military Training issued orders 'that the Army is to train for European Warfare next year (and [against] a best equipped enemy)'.[15] Despite its inability to persuade its political masters of the need to accept a continental commitment until 1939, the General Staff remained wedded to this priority, and it determined its instructions for the training of troops in the UK for the rest of the 1930s.[16] Thus, for example, in May 1936 Aldershot Command conducted a paper exercise to study how to handle a corps in a theatre like France, Eastern Command examined how the field force might be moved to France, and in 1937, Montgomery-Massingberd's successor, Sir Cyril Deverell, held a staff ride in Bedfordshire to practise an offensive by a force of one mobile and four infantry divisions attacking through the Aachen gap between Belgium and the Ruhr.[17] Deverell was dismissed in December 1937 because of his support for the policy of preparing the army for a continental land war, but his successor, Lord Gort, obstinately adhered to the same policy.[18]

However, it was one thing for the General Staff to see beyond the army's immediate commitments as a colonial police force and to identify in advance the kind of operation that it would have to undertake in the opening stages of the Second World War. It was quite another for it to be able to devise an appropriate doctrine to fight those battles, to impose a common under-standing of that doctrine on the army, and to ensure that troops actually trained on that doctrine.

Although much of the regular army spent most of the inter-war period acting as a colonial police force, it would be wrong to suppose that the General Staff focused its thinking about doctrine on colonial operations and neglected to think about the problems of a continental war, or that it only began to switch its mind to the continent in the late 1930s. Colonial soldiering did not encourage the General Staff to wait until the report of the Kirke Committee in 1932 to begin to digest the lessons of the First World War. The General Staff promulgated its initial thoughts on the lessons of the Great War as early as 1920, when it published the first post-war edition of the *Field Service Regulations*. Two amended editions followed in the 1920s, the second in 1924 and the third in 1929. It is testimony to the priority that the General Staff gave to thinking about 'big wars' that it did not publish a new edition of its doctrine for operations in aid of the civil power until 1923.[19]

By the end of the 1920s the General Staff had developed a combined arms doctrine that built on the experience of the First World War and on post-war exercises, and represented a sharp break with pre-1914 doctrine. Before 1914 the General Staff asserted that the infantry was the dominant arm, and that the other arms existed to pave the way for the decisive attack that the infantry would deliver with their bayonets.[20] By 1929 they recog-nised that victory would depend on operations in which all arms combined to achieve surprise, coupled with the deployment of a minimum of manpower and the maximum of mechanically generated firepower.[21] This was a major step towards a doctrine that promised to achieve victory on the battlefield at much less human cost than the army had suffered during the First World War.

But it would be an exaggeration to suggest that by the 1930s British doctrine was perfectly suited to the kinds of battles that the army engaged in during the first half of the Second World War. It was one thing for the General Staff to assert what field commanders and regimental officers should do. It was another for them to ensure that they did it. The General Staff's intellectual hold over the army remained almost as tenuous in the inter-war period as it had been before 1914.[22] This was not for want of trying. The Staff did try to impose their will on the army. In *FSR* (1920) they insisted that 'The army will be trained in peace and led in war in accordance with the doctrine contained in this volume'.[23] Successive Chiefs of the Imperial General Staff tried to ensure uniformity in training practices by disseminating reports on manoeuvres. Beginning in 1926, *Army Training Memoranda* were issued down to all officers of lieutenant colonel's rank.[24] But they were handicapped both by the nature of British doctrine, and by the weakness of the machinery at their disposal to oversee army training.

*FSR*s expounded general principles. For both cultural and practical reasons connected with the widely differing demands likely to confront troops, *FSR* deliberately did not prescribe how those principles should be applied. As *FSR* (1929) explained, the principles of war 'have a general and constant application to warfare. They can be defined, but their relative importance and the method of their application are constantly varying.'[25] It was left to senior officers, guided by their trained judgement and long experience, to determine for themselves how to apply them.[26] Colonial commitments undoubtedly worked to foist this laissez-faire policy onto the General Staff. To have gone further than espousing general principles would have run the risk that parts of the widely scattered army would have prepared for the wrong kind of operations. The application of any particular element of doctrine was likely to have to be very different on, for example, the Northwest Frontier of India, the Egyptian desert, or the plains of the Low Countries.

But there were also deeper cultural reasons why the General Staff espoused this attitude towards the definition and application of doctrine. This indulgent approach to the interpretation of doctrine was also the creation of a particular notion of what it meant to be 'British' that had developed since the eighteenth century. It was widely assumed that one of the factors that set the British apart from the Germans or French, and made them superior, was the fact that their actions were determined by 'character', not abstract reason and prescriptive rules. As *FSR* (1929) explained, 'The outstanding cause of this variation in the value and in the application of the principles of war is that, in war, the human factor predominates, and the human factor is a very varying quantity'.[27] A readiness to muddle through was a trait that was supposed to distinguish the British from their continental neighbours.[28] Unsurprisingly, the result was a good deal of confusion as to the exact status of the doctrine embodied in *FSR*. In 1923 Sir Philip Chetwode, the GOC-in-C Aldershot Command, told his subordinates that it was perfectly acceptable that 'individually we may differ as to its application. A common doctrine does not mean cramping initiative, it merely means team work as opposed to individual play.'[29] The tendency to 'bellyache', to regard orders as the basis for discussion, which was to cost the British so dearly in North Africa in 1941–2, therefore had deep roots in the pre-war British army.

Colonial commitments did not, therefore, prevent the General Staff from giving a high priority to instructing the army at home to train for a 'national war', nor did they dissuade them from devising a doctrine that was certainly better suited to such a war than the one they had tried to implement at the start of the First World War. Nor did they prevent them from devising training programmes for the army at home to prepare for a major European conflict. But the nature of that doctrine, with its emphasis on the application of principles determined by the judgement of individual commanders, was an obstacle to establishing a uniform training system, and was in part the result of the army's colonial commitments. Furthermore, it was one thing for the General Staff to lay down priorities and to set paper exercises. It was another for it to be able to impose its will on subordinate commanders, and for troops then to be able to carry out realistic training to practise that doctrine. And it was the inability of much of the army to do that which proved to be one of its most serious weaknesses in the opening stages of the Second World War.

The General Staff did not only have to overcome intellectual obstacles in establishing its ascendancy over the army. It also had to overcome more practical ones. Their attempts to impose a common understanding of doctrine were also impeded by a weak management and inspection system. The CIGS had so many other pressing duties that he had little time to oversee the higher training of the army. In practice it became the responsibility of individual GOC-in-Cs, each of whom had his own ideas about how his own troops should be trained.[30] Until 1939 the army's inspection system was too feeble to ensure that commanders were training their troops in accordance with official doctrine. Only the Royal Artillery and the Royal Tank Corps had inspectors who were unencumbered by other duties. The infantry had the Director of Military Training, and the cavalry had the Commandant of the Equitation School. But both of them, like the CIGS, had other time-consuming duties.[31] The most senior of the inspectors was only a major-general, which made his relations with GOC-in-Cs, who were either lieutenant-generals or full Generals, problematic. It was not until July 1939, when two very senior officers, Ironside and Kirke, assumed the duties of Inspector General of Overseas and Home Forces respectively, that the inspectorate finally acquired the authority it required, but by then it was too late.[32] Attempts to secure uniformity at a lower level also failed. Before going to command their unit aspiring unit commanders were supposed to attend the Senior Officers School, where it was hoped they would absorb a common tactical doctrine. But as the course lasted only for three months, it hardly gave them the opportunity to unlearn the sometimes idiosyncratic lessons they had been taught by their own commanders. Staff College graduates were also charged with the function of spreading a common doctrine throughout the army. But as few of them spent much time with their unit after graduating and before undertaking a series of staff jobs, they had too little intimate contact with the troops to do this. It was, therefore, hardly surprising that regimental idiosyncrasies could flourish, and that by 1927 the Director of Military Training was complaining that many regimental commanding officers were quietly flouting War Office training directives.[33]

But even a much more prescriptive and powerful inspection system would not have been able to overcome a whole series of mundane practical obstacles to realistic and uniform training on a common understanding of doctrine. The training of the regular army at home and overseas was organised according to an annual cycle that was divided into two parts. Individual training took place in the winter. It was followed by collective training in the spring and summer, culminating in the autumn in formation manoeuvres. In the first part of the cycle the emphasis was largely on training individual soldiers in skill-at-arms. In the second part collective training was designed to train commanders and units of different arms to work together and to impress on them the need for cooperation between all arms.[34] In theory this should have ensured that all troops and commanders received adequate training. In practice, it did not. This was partly because the way in which the training cycle allocated time was lop-sided. The time set aside for training individual soldiers in their basic skills was ample. Adequate time was found for sub-unit and unit training. But formation training – that is the training of brigades, divisions and corps – was crammed into a period of only about five weeks beginning in mid-August.

However, the failure of the training regime to produce units and formations capable of fighting a 'national war' also owed a good deal to the demands of colonial soldiering. Fighting 'small wars' and insurgencies too often took priority over training to fight a national war. Units actually manning colonial garrisons, about half the regular army at any one time, tended to concentrate on the immediate tasks of imperial policing rather than on preparing to fight a 'national war'.[35] In the opinion of Sir Edmund Ironside this meant that in the case of units in India, a British battalion that had spent three or four years in an isolated garrison on internal security duties 'is almost useless as a military force. It concentrates on turn-out, barrack-inspections and cook house sanitation, and shows an outward shell of great brilliance, fit to throw dust in the eyes of half our guard-mounting [? illegible] generals.'[36]

Even in the UK the practical priority that had to be afforded to on-going colonial commitments had a major impact on the way in which the regular army trained. The Cardwell system was designed to meet the needs of an army whose immediate commitment was to garrisioning the empire. That meant that units at home were maintained on a cadre basis. On mobilisation they could expect to be filled up with reservists recalled from civilian life. But reservists were not recalled to the colours for annual training, and in peacetime battalions in the UK were further depleted because they might lose as many as 40 per cent of their trained personnel each year as men were drafted overseas to maintain their linked battalions in the colonies at full strength.[37] Consequently, units in Britain were often too weak to perform realistic training exercises. In December 1920 Chetwode complained that training in Britain was in a state of 'suspended animation', not just because of shortages of money but also because so many troops had been sent to suppress rebellion in Ireland.[38] Sir Ivor Maxse, the GOC-in-C of Northern Command complained that, rather than training, 'We are employed, to a great extent, as policemen, housemaids, orderlies, gardeners and grooms.'[39] In 1921 the army was stretched so thin that the only forma-tion in the UK able to do any collective training was the Aldershot Experimental Brigade. By 1923 the standard of formation training was so backward that Cavan decided that it would be a waste of effort to carry out any collective training above brigade level.[40] Similar problems persisted into the 1930s. In 1937, for example, the collective training of both 1st and 4th divisions was interrupted when some of their units had to be rushed to Palestine.

Realistic training was also impeded by a shortage of training grounds. In Britain the Military Manoeuvres Act allowed the government to issue an Order-in-Council enabling troops to carry out exercises across a designated area. But it could train no more than once every five years in the same area, and then only under strict guidelines which robbed the manoeuvres of much of their realism. Soldiers were forbidden to enter almost any kind of building, including farmyards, orchards or enclosed woods, or to dig entrenchments if they risked damaging antiquities, woods or areas of outstanding natural beauty. The Act also ensured that the War Office would have to pay compensation to any property owners whose land, livestock or buildings were damaged.[41] Field training, therefore, included a large element of fantasy and make-believe. Roadblocks, craters and trenches were repre-sented by white tapes, anti-tank mines by bricks, and fire by thunder-flashes, flares, signalling lamps and blank ammunition. Mechanisation only made the problem worse because mechanised forces required even larger manoeuvre grounds than unmechanised ones. In 1933 Eastern Command, centred on Colchester, did not have a single training ground large enough to accommodate more than a brigade.[42] The British army only held two corps-level manoeuvres in Britain between the wars, the first in 1925 and the second in 1935.[43]

It might be thought that the wide-open spaces of the empire would have provided troops with the ground they needed, but in many cases it did not. In garrisons like Shanghai, Malta and Gibraltar, there just was not enough room for realistic unit and formation training. In India units deployed on internal security duties were usually accommodated in cantonments on the edge of large towns and cities. Many of them, like Meerut, had no nearby manoeuvre grounds.[44] Furthermore, during the hot season units stationed in the plains were usually divided and companies and squadrons rotated to distant hill stations. This gave officers and men some respite from the extreme heat, but meant that their field training had largely to be directed towards hill warfare. Egypt, however, was an exception to all this.

Commanders and troops stationed in Egypt had the advantage of training in units that were up to full strength, and they were able to train across the almost limitless land that they were to fight over in 1940. Therein lay one of the reasons why O'Connor's small force was able to defeat the much larger Italian 10th Army in the opening campaign of the desert war.

The material base

In the realm of material the army suffered from two major shortcomings in the opening years of the Second World War. It did not have enough of it, and some of the equipment that it did have was not designed for the job in hand. The first problem was not primarily caused by the army's commitment to colonial soldiering. Until 1939, successive governments had rejected the idea that they should prepare a large army to fight on the continent, and so they had made no preparations to provide it with the necessary quantities of equipment. In 1939 the British munitions industry consequently had little surge capacity, and did not expect to be able to produce enough equipment for the putative 55-division army until at least two years after the start of hostilities. The loss of the whole of the BEF's equipment at Dunkirk only served to add another year to this timetable.

Before the war the government's refusal to prepare an army for the continent meant that units in training often had to improvise equipment they did not have. Fifteen-hundredweight trucks painted grey were sometimes used to represent light tanks, infantry tanks were represented by trucks fitted with a conspicuous wooden 'T', and anti-tank guns by wooden mock-ups or flags.[45] Coloured flags sometimes represented whole sub-units of infantry, tanks and artillery.[46] In 1927, for example, the men of 2nd Battalion, Royal Fusiliers were introduced to 'The Rhine Army Tank [sic], a weird contraption consisting of a couple of bicycles with canvas round them'.[47]

However, the inter-war army's commitment to colonial campaigning did have a significant influence on the design of some of the weapons with which it tried to fight the Second World War. In 1918 the BEF had possessed a full outfit of equipment designed to fight slow-moving set-piece battles. Doctrinal developments after 1919 placed a premium on the restoration of mobility to the battlefield by all-arms cooperation and the generation of overwhelming firepower. To translate this doctrine into reality, the army needed to develop a new generation of weapons that combined weight of fire without sacrificing mobility. Ideally they also had to be mechanically reliable, cheap and easy to mass-produce, and supported by adequate spare parts and a maintenance system that could repair damaged equipment expeditiously.

In practice, the demands of colonial soldiering, which often required troops to operate at the end of long and tenuous supply lines, meant that the army placed more emphasis on lightness and mobility than it did on the ability to generate overwhelming firepower. It retained the Lee Enfield rifle and did little to develop an automatic rifle or sub-machine gun for fear that they would only encourage soldiers to waste precious ammunition. It was only when they confronted German troops armed with the MP38 (1938) that they realised their mistake and began to produce their own inferior counterpart, the Sten gun.[48] The Bren light-machine gun, which began to replace the First World War Lewis gun in 1937, was light and robust. But it was magazine, rather than belt fed, a design feature intended to economise on the expenditure of ammunition. That meant that it could not generate the same rate of firepower, or the same terrifying noise, as the German MG 34 or MG42. The 25-pdr field-gun that the army adopted in 1937 was reliable, robust, and could sustain a high rate of fire. It was an effective man-killer against troops in the open, and it could neutralise troops behind cover. But the quest for lightness and mobility meant that it sacrificed shell-weight.[49] However, there was one weapon that actually suffered from the fact that it was not designed with a colonial campaign in mind. The 2-pdr anti-tank gun was a perfectly adequate weapon when it was employed against the lightly armoured Panzer Is and IIs at the relatively short ranges that were possible in the close country of France and Belgium in 1940. But it failed against the more heavily armoured Panzer IIIs and IVs that were able to stand off and bombard it on the flatter and more distant battlefields of North Africa in 1941–2.[50]

The army's colonial commitments were not, however, responsible for the fact that it went to war without a sufficient number of reliable medium tanks. This major gap in its armoury owed everything to the Treasury's refusal in 1932 to continue funding a highly promising line of tank development. The Director of Mechanisation therefore had to order all development work to stop on a replacement for the army's existing medium tank. The result was that when rearmament began in 1936–7, the army did not have a satisfactory medium tank with a powerful purpose-built engine that it could put into production.[51] The Germans thereby established a lead in medium tank design that they never lost until the end of the war.[52]

By 1940 the British army's colonial legacy therefore meant that it was equipped with some weapons that were inferior to those of their main enemy. However, the influence of colonial campaigning was not entirely negative. The need to police a world-wide empire had long encouraged the military authorities to rely on high technology as a force multiplier. It was not accidental that the General Staff had decided on a gradual policy of mechanisation shortly after 1918, and that by 1939 the British possessed the first all-motorised army in the world. Mobility was important if the small British regular army was to police a large empire. Furthermore, colonial soldiering had taught the army one lesson that was to prove of crucial importance, and to give it a vitally important advantage. That was the over-riding necessity of ensuring that frontline troops had adequate logistical support. The importance of doing so was something they had learned over two centuries of colonial soldiering. Montgomery had often been criticised for placing too much emphasis on ensuring that his troops were properly supplied before attacking. Rommel, conversely, has been praised for taking risks with his logistics. This is illogical. In North Africa Rommel could gain tactical successes, but his lack of adequate transport meant that he could not translate them into operational victories.[53] The German army's willing-ness to close its eyes to logistical realities could be made to work across the short distances involved in fighting in France and Poland, and against enemies with an inferior operational doctrine, lacking sufficient air support, and handicapped by serious strategic errors. But in Russia, in North Africa, and finally in Normandy, the German effort foundered. And it did so in part at least, because their reach exceeded their logistical grasp, and because, unlike the British, they lacked sufficient motor transport successfully to cut loose from their railheads.

The personnel base – the regular army

The British army's personnel base between the wars was divided in two. The small, long-service professional army existed to fulfil all of the missions outlined above. The part-time volunteers of the Territorial Army had but a single role, to form the basis of a greatly expanded army in the event of a major national war.

One of the besetting problems that confronted the regular army was its inability to recruit up to its establishments, but the extent to which colonial service dissuaded potential recruits is difficult to determine. On the one hand the prospect of foreign travel was an inducement for some men to enlist.[54] On the other hand once the rank and file had been posted overseas they were not normally given home leave, and they could expect to spend up to six years in India before returning to the UK. By the mid-1930s the Adjutant General had identified the alarm of fathers, mothers and girl-friends at the prospect of a long separations from their sons and lovers as one reason why more men were not coming forward.[55] But there were plenty of other reasons why the army could not find the numbers of men that it sought that had nothing to do with its colonial commitments. Many recruits could not meet the physical standards set by the military authorities. In 1928, for example, 12.4 per cent of recruits in the depots were discharged due to physical or medical reasons.[56] Potential recruits were also perhaps dissuaded from presenting themselves because the positive public representation of the soldier prevalent during the First World War, as the heroic defender of hearth and home, faded quickly after 1918, to be replaced by the image of the regular soldier as a drunken libertine.[57] Some senior officers pointed to the prevailing atmosphere of anti-militarism.[58] Low pay and strict regimental discipline acted as further disincentives. Serving soldiers were not always a good advertisement for the army. And too many soldiers, after completing their period of colour service, found themselves cast onto the civilian job market without any useful skills.

Morale was a problem for the regular army in the inter-war period, particularly in units in Britain. It was degraded by many of the factors that impeded realistic training – the shortage of manpower in units, the shortage of modern equipment, and the shortage of adequate training grounds. As one senior officer wrote in 1936, troops taking part in exercises became disheartened. 'The men themselves enjoy real training for fighting, but they recognise as quickly as officers when the stage is reached where the imagination is overstretched.'[59] The training of the regular army in the inter-war period had entered into a downward spiral. Shortages of officers and men, with the colours, combined with a lack of modern arms and equipment, led to unrealistic training and, according to another senior officer, 'a general decline in tactical sense and a want of reality in all training'.[60] The result was a growing number of deserters and men seeking to be discharged on compassionate grounds in the mid-1930s.[61]

The personnel base – the Territorial Army

In the aftermath of the First World War the Lloyd George government and the General Staff had agreed that in the event of another great war the regular army would form merely the spearhead of Britain's effort on land.[62] They would not repeat the apparent mistake committed by Lord Kitchener in 1914–15 and try to raise New Armies from scratch. Instead Britain would look to the Territorial Army as the basis for the greatly expanded army that would be essential in the next 'national war'. But it was a role that the Territorials were both ill prepared and half-unwilling to fulfil, although for reasons which had nothing to do with the regular army's commitment to colonial soldiering. In the eyes of the General Staff the purpose of training in the Territorial Army in peacetime was to provide a framework for its expansion at the start of a national war. This meant that training had to be designed not only to foster *esprit de corps* and discipline in existing units. The Territorials also had to produce a sufficiently large cadre of instructors to train its newly raised second line units following the outbreak of war.[63] But these goals were incompatible. The former implied that men who trained together in peacetime would remain together after mobilisation and fight as a unit. But the General Staff's priority meant that on mobilisation Territorial units would be broken up to provide training cadres for wartime-recruited second line units. That was not welcomed by many Territorials. The integrity of their units was something they valued highly. In 1934, according to one Territorial battalion commander, few men would be attracted to enlist in peacetime if their only role was to become 'a training cadre constituted to expand the armed forces of the Crown on mobilization'.[64] Furthermore, when they were reconstituted in the early 1920s, as an economy measure their peacetime establishment was fixed at only 60 per cent of their wartime establishment. The result was that when the expansion of the Territorials began in 1939, units simply lacked the trained instructors within their own ranks that they needed if they were to throw off a training cadre to raise a duplicate unit and to maintain their own efficiency when they absorbed large numbers of new recruits to raise themselves to their wartime establishment.[65]

The efficiency and effectiveness of the Territorials as the basis for a greatly expanded wartime army were further degraded by their inability to achieve more than elementary levels of training. They confronted all of the practical problems that hampered the regular army plus other problems that were peculiarly their own. When the army began to receive the new equipment with which they were to fight the Second World War, the Territorials were inevitably last on the list, far behind regular units. Progressive training, even of single sections, was often difficult because of the rapid turn-over in personnel, and because only rarely did all the men of a section actually attend a drill meeting on the same evening.[66] The quality of local training facilities also varied greatly. Many units in rural areas were scattered across half a dozen stations and lacked adequate drill halls or parade grounds. It was, therefore, difficult to inculcate a common standard of efficiency and a common doctrine into the whole unit. Company and platoon commanders had different levels of enthusiasm and knowledge, and so when the whole unit came together it was often not easy for the commanding officer to predict how his subordinates would react.[67]

Territorial units and formations were in theory supposed to undergo tactical training at one of the four weekend camps each unit attended and the annual camp, which lasted for a fortnight. In reality combined arms training rarely took place even at the annual camp. So rudimentary was the state of individual training that many Territorial formations confined their work at camp to unit and sub-unit training.[68] In 1936 the War Office openly acknowledged that only if units were particularly efficient was collective training above unit level justified at the annual camp.[69] The only training most senior Territorial officers received in combined arms operations took the form of TEWTs (tactical exercises without troops), and consequently they rarely had 'the actual chance of seeing the co-operation of all arms'.[70] The result, according to one Territorial division CRA (Commander, Royal Artillery), was that by 1939 the Territorials knew even less of combined arms practices than the regulars. Their gunners had forgotten their *raison d'être* and the infantry thought only of the fire support they could generate within their own unit.[71]

Conclusion

By 1918 the BEF had come close to perfecting combined-arms operations on the battlefield. But as early as 1922 the Earl of Cavan recognised that the intimate all-arms cooperation that had characterised the operations of the army in the closing stages of the First World War was being lost.[72] Reflecting on his experience as Commander of 3rd Division in Britain in 1926, Sir Jock Burnett-Stuart highlighted some of the reasons for this. 'The Command of a Regular Division at home in the 1920s was not an exciting job; all units were below establishment, training areas were difficult to find, ammunition was scarce, and armament and equipment were already out of date.'[73] By the late 1930s many of the shortcomings in combined arms operations at the tactical and operational level that were to vitiate the performance of the army in the first half of the Second World War were deeply embedded. In 1937 the General Staff lamented, for example, that gunner officers lacked sufficient practice in the art of delivering fire support for the other arms.[74] Cooperation between infantry and tanks was similarly problematic. Brigadier A.G. Paterson, the commander of 1st Tank Brigade, commented that

> I have done far more work against real infantry than in the past and am frankly shocked at their ideas of anti-tank defence and standard of training. It is not the poor devils fault, they are short of equipment and for the most part have never seen tanks.[75]

It is too simple to lay the blame for these shortcomings solely on the army's commitment to colonial soldiering between the wars, but nor should that factor be ignored. After 1919 the General Staff's thinking about future doctrine was strongly oriented towards a continental commitment and fighting a 'national war'. However, the immediate and everyday reality of soldiering in the regular army was dictated by the needs of colonial garrisoning, and it created powerful countervailing forces that undermined the General Staff's ability to impose its vision of the future onto the army as a whole. The strongest of these was the pattern of deployment of the regular army dictated by the Cardwell system. About half of the regular army was, at any one time, scattered in usually small garrisons across the globe, and units in Britain were maintained at cadre strength to sustain them. This created a plethora of practical problem in training troops for a 'national war'. But equally important was the fact that the General Staff's intellectual control of the army was tenuous. This was only partly due to the requirements of colonial soldiering. General Staff doctrine enunciated principles, it did not lay down prescriptive drills, in part because British doctrine had to meet the widely differing requirements of soldiering in India, Africa and Europe. But not all the blame for this should be placed on the army's colonial configuration. The General Staff's unwillingness to be prescriptive and to dictate to subordinate commanders how they must behave was also the product of some profound cultural forces in British society, a peculiar sense of 'Britishness' that suggested that what made the British unique was their unwillingness, compared to their continental neighbours, to allow their actions to be governed by prescriptive ideas. To be British was to be pragmatic and to 'muddle through'.

But there were other reasons why the British Army was ill prepared for the war it had to fight in 1939 which had little to do with its colonial commitments. Not the least of these was that in the late 1930s the General Staff failed to impose its vision of the future role of the army onto its political masters. Industry was thus not prepared until it was too late to supply the army with the equipment and munitions it needed, and the Territorials were equally ill prepared to perform their allotted role as the basis for a greatly expanded national army. Together, these two factors contributed in large measure to the failures of the army between 1940 and 1942. However, they owed almost nothing to the demands of colonial soldiering between the wars. They owed almost everything to the reluctance of inter-war governments to accept the political case for a continental commitment and to spend the money that industry and the Territorials needed if they were to prepare themselves adequately for the task before them.

The fundamental reason why the army often performed poorly on the battlefield between 1940 and 1942 was rooted in the fact that its pre-war regular cadre was expanded haphazardly over tenfold between September 1939 and March 1941. Almost any organisation would have experienced problems in maintaining a high level of efficiency faced by such a challenge. Pre-war colonial commitments contributed to the fact that the regular army was not fully prepared to meet these challenges, but they were not the fundamental cause of them. It is, therefore, perhaps apt to conclude on a culinary note. The record of the preparations of the inter-war army for a major continental war contained a few plums. It also contained too much duff.

Notes

1 Public Record Office (PRO) WO 279/59. War Office Exercise No. 2. Winchester, 9 to 12 May 1927. See also PRO WO 32/2382. Memorandum on Army Training Collective Training period 1928, 26 Nov. 1928.
2 War Office, *Field Service Regulations. Vol. 1. (Organization and Administration), 1930*, London: War Office, 1930, 1–2. (Henceforward *FSR.*)
3 War Office, *FSR Vol. II (Operations)*, London: War Office, 1929, 1.
4 The best accounts of these operations are A. Clayton, *The British Empire as Superpower 1919–1939*, London: Macmillan, 1986 and Major-General Sir C. W. Gwynn, *Imperial Policing*, London: Macmillan, 1934.

5 Ministry of Defence, *Joint Warfare Publications, British Defence Doctrine* London: MOD, 1996, 3.9 –3.11.
6 War Office, *Training Regulations, 1934*, London: War Office, 1934, 1.
7 Lidle Hart Centre for Military Archives (LHCMA). Kirke mss I/14. Kirke, The Experimental Brigade. Lecture by Colonel W.M. St G. Kirke, n.d. but *c.* late 1922.
8 PRO WO 279/54. Report on Staff Exercise held by the CIGS, 30 October to 3 November 1922.
9 PRO WO 279/55. Report on Staff Exercise held by the CIGS, 9–13 April 1923.
10 PRO WO 279/60. Report on the Staff Conference held at the Staff College, Camberley, 16–19 January 1928.
11 LHCMA. Liddell Hart mss 1/322/34. Gort to Liddell Hart, 28 July 1931, and 1/322/25. Gort to Liddell Hart, 25 May 1932.
12 Captain J.R. Kennedy, 'Army Training', *Journal of the Royal United Services Institute*, vol. 77 (1932), 714.
13 PRO WO 32/3115. Milne, *Army Training Memorandum Number 4A. Guide for Commanders of Regular Troops at Home*, 1932, 29 Dec. 1931.
14 PRO AIR 2/1664. Slessor to Peck, 18 June 1934; R.H. Larson, *The British Army and the Theory of Armoured Warfare 1918–1940*, Newark: University of Delaware Press, 1984, 176; PRO WO 32/2847. Minute, Montgomery-Massingberd to Military Members of Army Council, 15 Oct. 1934.
15 LHCMA. Liddell Hart mss 11/1934/48. Talk with Colonel G. Le Q. Martel, 9 Sept. 1934; PRO WO 277/8. Anon., *Fighting, Support and Transport Vehicles and the War Office Organization for their Provision* (War Office, 1948), 2.
16 See, for example, LHCMA. Liddell Hart mss 11/1936/99. Talk with Field Marshal Sir C. Deverell, 12 Nov. 1936.
17 LHCMA. Liddell Hart mss 11/1935/142. War Office Exercises without troops and war games, 1935–36; B. Bond, *British Military Policy between the two World Wars*, Oxford: Clarendon Press, 1980, 252.
18 J. P. Harris, 'The British General Staff and the coming of war, 1933–1939', in D. French and B. Holden Reid (eds) *The British General Staff. Reform and Innovation, 1890–1939*, London: Frank Cass, 2002, 186–7; PRO WO 32/10326. Gort to Coordinating Committee of Army Council, 18 July 1938.
19 War Office, *Duties in Aid of the Civil Power*, London: War Office, 1923. I am most grateful to Mr Simeon Shoul for this reference.
20 *FSR* (1909), p. 20.
21 D. French, 'Doctrine and organisation in the British army, 1919–32', *Historical Journal*, vol. 44 (2001), 497–508.
22 A problem highlighted in the context of the army before 1914 in H. Strachan, 'The British army, its General Staff and the Continental Commitment 1904–14', in French and Holden Reid (eds) *The British General Staff*, 90–2.
23 *FSR* (1920), 13.
24 PRO WO 279/65. Report on the Staff Conference held at the Staff College, Camberley, 14–17 Jan. 1929. Comments by Milne and the DMT.
25 *FSR* (1929), 7.
26 Sir Ivor Maxse, 'Foreword', in Captain B.H. Liddell Hart, *A Science of Infantry Tactics Simplified*, London: William Clowes, 1926, vi.
27 *FSR* (1929), 7.
28 R. Colls, 'Englishness and political culture' in R. Colls and P. Dodd (eds) *Englishness. Politics and Culture 1880–1920*, London, 1986, 31; S. Collini, *Public Moralists. Political Thought and Intellectual Life in Britain 1850–1930*, Oxford, 1991, 323–41; T. W. Heyck, 'Myths and meanings of intellectuals in twentieth century British national identity', *Journal of British Studies*, 37 (1998), 192–99.

29 LHCMA. Liddell Hart mss 15/8/56. Some Remarks on Training made by Lt. Gen. Sir Philip Chetwode to Aldershot Command on 21 April 1923. Issued by General Staff, Aldershot, May 1923.

30 Col. R. Macleod and D. Kelly (eds) *The Ironside Diaries 1937–40*, London: Constable, 1962, 391–2.

31 War Office, *King's Regulations for the Army and the Army Reserve 1928*, London: HMSO, 1928, 2–3.

32 Macleod and Kelly (eds) *The Ironside Diaries*, 76.

33 PRO WO 279/57. Report on the Staff Conference held at the Staff College, Camberley, 17–20 January 1927 under the orders and direction of the CIGS.

34 General Staff, *Training Regulations*, 8–10.

35 T.R. Moreman,, ' "Small wars" and "imperial policing": the British Army and the theory and practice of colonial warfare in the British Empire, 1919–39', *Journal of Strategic Studies*, 19 (1996), 113–14.

36 LHCMA. Liddell Hart mss 1/401. Ironside to Liddell Hart, 29 May 1929.

37 PRO WO 279/60. Report on the Staff Conference held at the Staff College, Camberley, 16–19 Jan. 1928. Remarks by Major General W. H. Bartholomew, the Director of Recruiting and Organization at the War Office.

38 LHCMA. Montgomery-Massingberd mss 122/1. Chetwode to Montgomery-Massingberd, 30 Dec. 1920.

39 Maxse made these comments before an audience at the Royal United Services Institute who had listened to a lecture by Liddell Hart on infantry training. See B.H. Liddell Hart, ' "The man in the dark" theory of infantry training', *Journal of the Royal United Services Institute*, 64 (1921), 22.

40 LHCMA. Liddell Hart mss 15/8/56. Some Remarks on Training Made, 21 April 1923.

41 General Staff, *Training Regulations*, 118–28.

42 LHCMA. Aston mss. *The Times*, 22 April 1933.

43 PRO WO 279/56. Report on Army Manoeuvres, 1925; PRO WO 279/76. Report on Army Manoeuvres, 1935.

44 LHCMA. Liddell Hart mss 1/401. Ironside to Liddell Hart, 26 Jan. 1929.

45 LHCMA. Liddell Hart mss 15/8/93. Major General A.E. MacNamara, *Instructions Regarding Training in the Manoeuvre Area for 1935*, London: War Office, 1935; LHCMA. Liddell Hart mss, 15/3/23. General Staff, *Army Training Memorandum No. 20*, April, 1938, London: War Office, 1938.

46 General Staff, *Training Regulations*, 68.

47 *Royal Fusiliers Chronicle*, no. 19 (1927), 16.

48 I.V. Hogg, *The Encyclopaedia of Infantry Weapons of World War II*, London: Bison Books, 1977, 52–3; PRO WO 106/1775. Report of the Bartholomew Committee, n.d. but *c.* 2 July 1940.

49 PRO WO 277/5. Pemberton, The Development of Artillery Tactics and Equipment, 12.

50 PRO WO 277/5. Pemberton, Artillery Tactics and Equipment, 16, 88; B.T. White, *Tanks and Other Armoured Vehicles of World War Two*, London: Perage Books, 1972, 153–4; PRO WO 106/1775. Report of the Bartholomew Committee, n.d. but *c.* 2 July 1940; PRO WO 32/9642. Martel to Dewing and enc., 9 July 1940; PRO WO 199/3186. Report on the organisation and equipment of 1st Armoured Division, 26 June 1940; PRO WO 201/2586. Middle East Training Pamphlet No. 10. Lessons of Cyrenaica Campaign, December 1940–February 1941.

51 PRO WO 32/4441.CIGS to Secretary of State, 9 Oct. 1936.

52 J.P. Harris, 'British armour and rearmament in the 1930s', *Journal of Strategic Studies*, 11 (1988), 220–44.

53 M. Van Creveld, *Supplying War. Logistics from Wallenstein to Patton*, Cambridge: Cambridge University Press, 1977, 142–201; L.H. Addington, *The Blitzkrieg Era*

and the German General Staff, 1865–1941, New Jersey: Rutgers University Press, 1971, 159–76.

54 A. Dixon, *Tinned Soldier. A Personal Record, 1919–26*, London: The Right Book Club, 1941, 28–9.

55 PRO WO 32/2984. Adjutant General to Army Council, 15 Sept. 1936.

56 PRO WO 32/4643. Maj. W.A.S. Turner to Maj. C. W. Baker, 6 April 1936.

57 Colonel W.N. Nicholson, *Behind the Lines. An Account of Administrative Staffwork in the British Army, 1914–1918*, London: The Strong Oak Press and Tom Donovan, n.d.but first published 1939, 271.

58 PRO WO 32/2984. Burnett-Stuart to PUS, War Office, 22 Feb. 1936.

59 PRO WO 32/2984. Finlayson to GOC-in-C Southern Command, 15 Jan. 1936.

60 LHCMA. Burnett-Stuart mss 3. Burnett-Stuart, Southern Command. Annual Report on training of the Regular Army, 1936/37, Nov. 1937.

61 PRO WO 32/2984. Quarter-Master General to Adjutant General, 8 Nov. 1935; Adjutant General to Duff Cooper, 10 July 1936.

62 P. Dennis, *The Territorial Army 1907–1940*, London: Royal Historical Society, 1987.

63 PRO WO279/75. Report on the Staff Conference held at the Staff College, Camberley, 8–12 January 1934. Comment by Montgomery-Massingberd.

64 PRO WO279/75. Report on the Staff Conference held at the Staff College, Camberley, 8–12 January 1934. Comment by Lt. Col. Tozer.

65 Lt. Col. M.C. Eden, 'The organization and training of Territorial Army units', *Journal of the Royal United Services Institute*, 84, (1939), 135–43.

66 LHCMA. Liddell Hart mss 1/519/2. Montgomery to Liddell Hart, 16 July 1924.

67 Capt. P.A. Hall, 'The training of junior leaders in a county Territorial battalion', *Journal of the Royal United Services Institute*, 77 (1932), 586–8.

68 PRO WO279/75. Report on the Staff Conference held at the Staff College, Camberley, 8–12 January 1934. Maj.-Gen. E.F. Marshall.

69 War Office, Regulations for the Territorial Army (including the Territorial Army Reserve) and for County Associations, 1936, London: War Office, 1936, 141.

70 PRO WO 279/70. Report on the Staff Conference held at the Staff College, Camberley, 13–16 Jan. 1930. Comments by Major R.H. Lorrie.

71 Brigadier R.G. Cherry, 'Territorial Army Staffs and Training', *Journal of the Royal United Services Institute*, 84 (1939), 551.

72 PRO WO 279/55. Report on the Staff Exercise held by the CIGS, 9–13 April 1922.

73 LHCMA. Burnett-Stuart mss 6. Burnett-Stuart, Memoirs, chapter XVI.

74 LHCMA. Liddell Hart mss, 15/3/23. General Staff, *Army Training Memorandum No. 18, April 1937* (London: War Office, 1937).

75 LHCMA. Liddell Hart mss, 15/3/40. Paterson to Liddell Hart, 18 Sept. 1937.

3 Learning new lessons

The British Army and the strategic debate, 1945–50

Paul Cornish

Introduction

The British Army's transition from the high intensity of the Second World War (in Europe and elsewhere) to the very different strategic environment of the Cold War, in which the commitment to the occupation and defence of Germany took on increasing importance, has for long attracted the interest of historians and practitioners alike. With a strategic commitment to Europe a common factor, the working assumption has been that doctrinal and practical linkages were made (or should have been made) between these two strategic environments, with lessons learned in one rich experience being applied in the other. This assumed relationship between World War and Cold War has generated several lines of inquiry. After 1945, was the Army thinking and preparing for 'small wars' or 'major wars'? And if the latter, what operational and tactical lessons had been learned, particularly from the intensive joint and combined operations in Europe? At the military strategic level, with all that it had learned from 1939–45, what did the British Army see as its purpose and its preferred *modus operandi* after 1945? Where was the Army's Cold War focus: Europe; the Middle East; the Far East; elsewhere?

Others have taken a more critical approach, exploring and explaining the disconnections between the two strategic experiences. In his study of the development of manoeuvrist thinking in the British Army after 1945, Kiszely laments the fact that, in spite of some British demonstrations of manoeuvrism during the war, and in spite of the Army's experience and understanding of the effectiveness of German manoeuvrism, the attritionist style which had dominated the Second World War proved difficult to escape after the war: 'It was perhaps surprising … that after 1945 the Army did not itself seek to adopt a less attritional, more manoeuvrist approach – but there is little evidence that it did so'.[1] Kiszely offers a number of reasons for the failure to learn the manoeuvrist lesson. First, the Army had been on the winning side. As with any organisation, success is not often followed by a full-scale, root-and-branch review of operating procedures. (It might be added that success is even less likely to be followed by the adoption of the procedures of the losing side.) Second, there were many in the higher echelons of the Army who were quite convinced that manoeuvrism had been a dangerous dallying with fashion, and that the highly centralised, attritionist approach had worked. Most prominent among the advocates of attrition was, of course, Montgomery, whose influence on the post-war Army was considerable:

> As Chief of the Imperial General Staff (CIGS) in the immediate post-war years, Montgomery's view, not unnaturally, carried considerable weight, and his admirers occupied influential posts in the Army for many years after the war. Staff College instruction at the time, unsurprisingly, emphasised a highly controlled style of warfare which sought to impose order on the battlefield, moving firepower to destroy enemy strengths, rather than a style aimed at flourishing in situations of chaos and uncertainty.[2]

Finally, Kiszely argues that the attritional approach was more suited to the evolving role of the British Army of the Rhine (BAOR): 'A large proportion of the army settled down in West Germany to confront the Russian, and subsequently Warsaw Pact forces, limited by constraints which allowed little room for manoeuvre.'[3]

This chapter offers a rather different account of the British Army's transition from World War to Cold War. Of course there were many campaign reviews and what might now be called 'lessons learned' processes, often in the form of well organised battlefield tours, the papers accompanying which remain a valuable historical resource.[4] In terms of training and doctrine, the value of the recent campaign in Europe, in particular, would have been obvious to all. But the contention of this chapter is that the search for either connections or disconnections between the two strategic experiences – World War and Cold War – misses the point. At least until the outbreak of the Korean War in 1950, there was no transition, and no application of lessons learned from the Second World War to the new, evolving environment. During the formative period of the Cold War, the British Army's experience of the Second World War (in Europe and elsewhere) was neither applied nor overlooked; it was simply irrelevant. The British Army spent its time at best jockeying for influence, and at worst fighting for survival, as the battle lines of the Cold War were being drawn. The Army, in short, was preoccupied with developing a strategy for the inter-service war in Whitehall, and its best and brightest were preoccupied, not with drawing and applying the operational and tactical lessons from the Second World War, but with the more pressing political need to secure for the Army a leading role – whatever that meant – in the evolving Cold War. The post-war strategic debate, and the Army's positioning within that debate, can best be understood by examining two closely related issues: threat perceptions and the allocation of resources.[5]

Threat perceptions

In the final months of the war the main, short-term danger was that of a less than complete victory for the Allies. It was feared that the war might end 'untidily', with pockets of resistance remaining and ignoring the call to surrender.[6] The Chiefs of Staff (COS) were also, even before Germany's defeat, alive to the danger of German resurgence.[7] Generally speaking, however, the German problem was thought to be temporary and easily manageable with the military resources available. The COS were sensitive to other, emerging threats, and since autumn 1944 had been aware that principal among these was, of course, the Soviet Union. By January 1945 the COS were in no doubt that the Soviet Union's intentions in Germany and elsewhere were at least uncertain and at worst inimical. The minutes of the 11th Meeting of 10 January 1945 contain numerous references to the 'Soviet threat' and the 'hostile intentions' of the USSR.[8] Reference was also made at this meeting to the danger of a resurgent Germany forming an alliance with the Soviet Union. In the months after Germany's defeat, however, the COS returned to the Soviet threat. It was felt that examination of military commitments for the period 1946–50 should be carried out with clear reference to 'the possibility of having to meet Russian aggression'.[9] For the COS and their staffs, the Soviet Union was now, unequivocally, the main threat to Europe and Germany, and therefore to the British forces on occupation duties. That said, it was believed that Germany was less likely to be the scene of conflict with the Soviet Union than the Middle East. Not only was the Soviet Union considered vulnerable in this area, with the oil-fields of Soviet Central Asia, but British possessions, resources and lines of communication were themselves seen to be most in need of protection.

During 1946, where Germany was concerned the threat was increasingly being seen as a matter of internal disorder, or of disaffection breeding communism, rather than a matter of a direct military, or even political, challenge. In mid-February the COS discussed the implications of likely cuts in the manpower and funding of the forces. One possible implication was 'a potentially dangerous weakness in Germany'.[10] If the fragile law and order structure imposed on occupied Germany were to break down, there might then be an unwelcome requirement for a disproportionate and costly commitment of troops to an area which, after the defeat of Germany, was a second-order problem. As time wore on the realisation also dawned that disaffection resulting from appalling economic conditions in Germany could give rise to a more serious security problem, whereby Germany would become a 'breeding ground' for communist doctrine exported from the Soviet Union or, perhaps, some sort of German nationalist revival. Beyond this, the COS seem to have accepted by 1946 that Germany by itself was no longer a concrete, active threat to British interests.

The Soviet Union remained a problem of a very different order. Comments made throughout the year, concerning a variety of British overseas interests, show that the Soviet Union now had the unrivalled lead in the

British military demonology. Military bases would be needed in the Middle East, partly for the protection of British interests there, partly for the security of sea lines of communication (SLOC) in the Mediterranean area, but also in order to provide air bases for a strategic bombing attack on the Soviet Union.[11] Soviet antagonism, potential or actual, was perceived in many different areas, demonstrating the breadth as well as depth of COS convictions. Well before the Soviet atomic bomb test (August 1949), Montgomery demanded that the scenario for an Army staff exercise being planned for May 1947 in Camberley include the use or threatened use of atomic bombs in Europe by an enemy. There can be little doubt which enemy he had in mind.[12]

The position at the end of 1946 was that the Soviet Union was perceived by the British military to be the most, or indeed the only, serious threat to Britain and her interests. This position had been reached slowly and cautiously. To suggest that the Soviet threat replaced the German, implying a crisp, coherent change of emphasis, is to simplify the issue. For the switch from Germany to the Soviet Union to take place, two judgements were required, both of which took time to develop. First, the COS and their planning staffs had to be confident that there could be no chance of German revanchism and that the 'German problem' was either one merely of maintaining law and order in the British zone of occupation, or simply symptomatic of the wider problem of creeping communism which would have to be addressed elsewhere. Second, the COS had to come to the conclusion that the Soviet Union was engaged in policies which could lead to war with Britain either intentionally or, more likely, through miscalculation. In April 1946, Alanbrooke, coming towards the end of his long appointment as Chief of the Imperial General Staff (CIGS), summed up this important realignment of perceptions. Germany, he felt, should not be considered a threat

> unless she was forced into co-operation with the Russians. Our major policy should then be to build her up sufficiently to obviate becoming a liability to ourselves, yet independent of Russian threats. ... Our long term policy must take full account of the fact that Russia is a more dangerous potential enemy than Germany.[13]

The German threat, such as it was, continued to be one of a breakdown in the internal security of the British zone. As in the past, however, this problem was addressed largely as a police matter, and was not the stuff of strategic planning.

May 1947 saw the production of the Future Defence Policy (FDP), indicating that military thinking *vis-à-vis* the German threat was to 'double hat' the inchoate anti-Soviet strategy.[14] The FDP report spoke in clear terms of 'the possibility of war with Russia', of 'standing up to Russia', of 'Russian expansion' and so on, all sentiments reflected in the COS papers, and in their thinking throughout the year. On New Year's Day 1947, Slim, then commandant of the Imperial Defence College, presented the IDC's general policy review. The COS expressed 'general agreement' with the review which identified Russia as the 'only potential aggressor', the 'only possible enemy', and cautioned in particular against any policy of appeasement.[15] The IDC review seems also to have nudged the COS towards the development of a characteristically British version of containment, one which would have a pronounced military component, set alongside political and economic measures.[16] That said, this was not to be territorially defined containment, requiring Germany to be garrisoned and defended: Germany was not at this stage seen as much of a bulwark against the might of the Russians. Instead, the IDC review argued that 'defence by the shield', against a Russian attack on Britain, 'would not be effective' and that the only viable policy would be, in effect, an early version of massive retaliation:

> Real defence in the atomic age must lie in making an aggressor realise that while he may wreck the nation he attacks, he will at once be subjected himself to such a counter blow that, whatever success he gains in his initial attack, will be purchased too dearly. Our main defence should be a counter blow the moment we are attacked, and the knowledge that it exists.[17]

The counter-offensive force was to be some fifty squadrons of twenty-five aircraft each. Military containment of the Soviet threat was to be achieved, in other words, by deterrence rather than by some territorial strategy, still less one involving Germany. The COS had been well aware, throughout 1946, that the Soviet Union represented a 'more dangerous potential enemy than Germany',[18] but the deduction was that Britain should shore up her defences in the Middle East, not Germany. Germany was still, largely, an internal security problem and was not to involve a major, strategic commitment of forces.[19] Even Montgomery, who might be expected to have grasped any opportunity to prove the Army's indispensability, found himself agreeing, in mid-summer 1946, that the occupation force should be reduced to just over four divisions; all that would be needed to meet basic occupation tasks.[20] It is clear, therefore, that by the middle of 1946, the British had decidedly not embarked on a policy to contain the Soviet Union with British armed forces deployed in Germany. Nevertheless, the British occupation zone in Germany did occupy a significant place in the evolving strategy of atomic deterrence. Although Germany was not one of the 'three pillars' of this new policy (defence of the UK base, defence of SLOC and defence of the Middle East), the defence of Germany was incorporated into the new policy simply because it was considered essential to resist territorial expansion of the Soviet Union. Hence, one aim of the evolving strategy of atomic deterrence was to 'ensure that Germany does not become a Russian Satellite'.[21]

The Chiefs of Staff were concerned, of course, with the possibility of an attack on Britain. In their discussion of the Imperial Defence College review on 1 January 1947, the COS were presented with three possible forms of attack: air attack; advance over land to within rocket strike range; and attack by submarines on the SLOC. Air attack would be by rocket or manned aircraft, with atomic, chemical or biological weapons. The COS were warned of the damage which might result from such an attack, with some 60 per cent of the 'white manpower' and industrial strength of the Commonwealth concentrated in Britain. The deduction was that the battle should be fought as far away (east) as possible. This in turn meant attacking the Soviet Union from the Middle East and destroying their supplies of oil. It also meant preventing enemy bases being established in Europe from which an attack could be launched.[22] Finally, Britain's perceived vulnerability demanded that air defences be brought up to standard. Attack over land to within rocket strike range was simply an extension of the first, and prompted similar conclusions. The COS were thus well aware of the vulnerability of Britain to air and rocket attack. Yet they were sanguine about the inability (for ten years to come) of Britain's air defences to come up to the mark. Once again, this can only be explained by faith in the deterrent effect of the West's atomic advantage, and particularly of the threat of a massive response.

If the main threat to Britain was from the air, and if air power would provide the best defence, there were clear implications for Britain's armed services, particularly the Army. It was not envisaged that the Army would provide any sort of land defence against the Soviets in Europe. No commitment to the defence of Germany was considered at this stage – no continental territorial strategy – even though the COS saw the need to ensure that Germany would not become a launch pad or airfield for the Russians: a paradox which sums up perfectly the tensions within Britain's evolving strategy for Europe. In late February 1947 the COS were made aware of the Joint Planning Staff view that manpower difficulties would require a run-down in the Army's deployed strength in Germany to approximately 30,000–40,000 by early 1949 and that occupation forces would be dispensed with altogether in 'Phase 2'.[23] That the Army was being forced to look elsewhere to justify its existence is evidenced by the minutes of two meetings held in December 1947.[24] Here, the Army fought to secure for itself a larger slice of the defence estimates cake by stressing (with the support of the Joint Intelligence Committee (JIC)) the possibility of air and sea invasion of the UK. Only the Army could deal with such a threat. The Royal Navy and RAF chiefs played this down: if threat and response were defined in air/maritime terms then there would be more resources for their two services. Against the weight of contemporary strategic orthodoxy, it is not surprising that the Army pinned its colours to the strongest post it could find. Static defence of Germany was not that post: a large-scale commitment of fighting forces would have required the Army's place in British strategy to be on a par with or higher than that of the Royal Navy or Royal

Air Force. But no member of the COS – not even the CIGS – was yet willing to make that argument.

From early in 1948 there is ample evidence that the COS and their planning and assessment staffs were becoming far more interested in Soviet military capabilities, atomic and conventional. In his memorandum on 'Future Strategy' of 7 February, Montgomery provided detailed information on the size and character of the Soviet threat. Montgomery described a Soviet build-up to a three-pronged attack into northern and central Germany, naming key towns and regions. The Soviets would use twenty-one divisions which would take between 6 and 15 days to arrive on the Rhine. By D+28 days, the Soviets could have seventy-three divisions on the Rhine.[25] A paper presented by Tedder, Chief of the Air Staff, referred to a 'conservative assessment' made by the Intelligence Staff that considerably more than 5,000 Russian aircraft would be made available in support of their 'field campaign'.[26] Yet, in spite of their evident awareness of Soviet military capabilities in Europe, and of the possibility of an accidental war, the British military did not as yet feel any need to take dramatic decisions or indeed to change their basic planning assumptions. A planning date of 1956–7 gave a decent interval for preparation while a conflict which broke out in the short-term would, by its very nature, be a surprise and therefore impossible to plan for. In any case, if the Soviets were to launch their massive forces at Western Europe before 1956–7, there would be little the West's divisions could do about it. This had immediate implications for Britain's military involvement in Germany. At the end of April, the Foreign Secretary Ernest Bevin wrote to his colleague A.V. Alexander, Minister of Defence, and expressed his concern over the Soviet threat to the security of the western zones of occupation.[27] Bevin nevertheless saw the danger of Soviet attempts to ruin the European Recovery Programme (ERP) by subversion, civil disturbances and so on. Their objective, he believed, would be 'to force the withdrawal of the allied forces from Germany and Austria, since those forces represent the real bar to the further progress of Communism westward'. Bevin assumed that the job of the forces in Germany and Austria was to 'fight', not an open conflict with the Soviets, which he dismissed as unlikely, but a political or ideological conflict. The forces of occupation would have to resist, not tanks, but civil disobedience, sabotage and 'local diversionary attacks'. By carrying out these tasks, and simply by their presence in Germany and Austria, the occupation forces could 'contain' the threat, as it was then defined. This provides another dimension to the notion of military containment examined earlier: what was now being proposed was a mid-point between 'passive' and 'active' containment, whereby the political or ideological threat could be contained in Germany by military means, albeit still without a territorial strategy as such.

Sir Brian Robertson's[28] response to Bevin's note was forwarded to the COS in late June.[29] Robertson concurred with Bevin; he anticipated the Soviets applying 'progressive' pressure to bring about the collapse of the ERP. This might involve infiltration of the trade unions and the police forces by way of 'softening up'. The object would be to undermine German confidence in the occupation regimes and in the German institutions which had so far been established. There would then be sabotage, aimed at the police, industry and at the 'mobility of the Armed Forces'. Bands of German guerrillas, sponsored by the Soviets, might make incursions into the western zones, with the aim of making the occupying powers' position untenable. Robertson recommended a resolute response to nip the problem in the bud – on political, economic and military levels. Like Bevin, he did not believe that the Soviets would attack. If they did attack, however, the sequence of events would not alter, it would simply be more rapid. In something of a giveaway statement, the occupying forces would then be under immense pressure as they tried to hold their ground: 'In that case it would be necessary to reconsider the use of Germany as a training ground for national service men, and the replacement of the present recruits by seasoned troops.' When the COS later discussed Robertson's letter, more of the Army's situation in Germany was revealed.[30] Sir Gerald Templer, the Army's Director of Military Intelligence, noted that Robertson's interest in 'seasoned troops' was inappropriate since 'there were no seasoned troops available'.

The Chiefs of Staff believed the Russians to be capable of launching a massive attack westward. The scale and direction of any Soviet attack was, however, the cause of some disagreement. In November, the COS heard that the Directors of Intelligence estimated that the Soviets could already have 174 divisions at 70 per cent strength, with all at 100 per cent strength after a further twelve months.[31] It was a 'practical probability' that all these divisions could be mobilised in five days.[32] The COS approved the intelligence paper – JIC (48) 100 – as a background for further planning and study.[33] Some days later another JIC assessment of the 'possibility of war before a certain date' was discussed. In discussion, the COS agreed to amend the JIC report (JIC (48) 121 (Final)) as follows:

> The Soviet land forces, with their close support aircraft are at present sufficiently strong to achieve extensive initial success against any likely combination of opposing land forces. They have also, at present, relative superiority in the performance of their armoured fighting vehicles and of certain other weapons; if war is delayed the maintenance of this superiority will involve them in a re-equipment programme to the detriment of other industrial projects. ... The Soviet leaders may consider, therefore that in view of the defensive preparations of the Western Powers, this superiority is dwindling and should be exploited without delay.[34]

The first full, intelligence assessment for 1949 appeared at the end of January.[35] Although the Soviets could start a land war 'at any time', before 1956–60 any early numerical advantages would be offset by the incompletion of the Soviet industrial plan. What was even more likely to hold the Soviets back, however, was – once again – the Western preponderance in the strategic air balance. It was thought that the Allies had the advantage in strategic bombing technology and techniques and that Soviet air defences were inadequate. To embark upon a land campaign, with at best only a temporary numerical advantage on the ground and while the air power umbrella above them was so heavily perforated, would have entailed immense tactical and strategic risks, and would have subjected the Soviet Union to a ferocious strategic air response.

A discussion in mid-April 1949 of the logistic aspects of Plan 'Speedway' – a plan to evacuate Europe in the event of Soviet aggression – provides another glimpse of the progress of threat assessments. According to a JPS report, Russian military aims were the destruction or neutralisation of Allied forces on the European and Asian mainlands, the seizure or neutralisation of Britain, the seizure of the Middle East and its oil resources, the seizure or neutralisation of all Allied air offensive bases, and, finally, the disruption of Allied SLOC by submarine interdiction, mining and air operations.[36] The JPS estimated that France and the Low Countries could be overrun within two months and that appropriate Channel and North Sea ports might only be available to the Allies for the first two weeks of a war. There was a threat of strategic bombing, but the authors of the report felt that, given the relative inexperience of the Russians in this form of warfare, this threat would be at its greatest only if the Soviets were able to establish themselves in Western Europe, i.e. within closer range of the UK. This, of course, was the basic rationale behind Britain's interest in defending Europe as far to the east as possible.

During the early months of 1950, before the outbreak of the Korean War, the idea of defending and even rearming Germany in the face of the Soviet threat had a surprising effect on British military threat perceptions in restoring life – albeit briefly – to fears which had been moribund, or at least peripheral, for some time: the dangers of German revanchism and, worse still, of a new 'Rapallo' agreement between Germany and the Soviet Union.[37] At this stage, the military staffs were showing a good deal more interest than the Foreign Office in the possibility of rearming Germany in order to boost the Western Alliance's defences in Europe. Part of this interest was no doubt a result of the awareness in British and American military circles (several months before the outbreak of the Korean War) that, in the event of war in Europe, the Western powers would probably be unable to offer any meaningful resistance.

In mid-March 1950, while discussing 'the arguments in favour of promising United Kingdom land reinforcements for the defence of Western Europe in the event of war',[38] two milestones emerge in COS/JPS thinking. First, the JPS had by now come to the firm conclusion that successful defence of the United Kingdom would not be possible in the event of a Soviet invasion and occupation of Western Europe: 'We conclude that the defence of the United Kingdom and Western Europe must be considered together. The foundation of the defence of Western Europe in war is a strong France.'[39] There was nothing novel in this observation:

British military thinking, both before and after the advent of the bomber and the rocket, had tended to see Northwest Europe as something of a buffer-zone or glacis. What makes this comment so interesting is that the JPS were now willing to declare their sense of Britain's dependence (no longer merely interest) on the security of Western Europe: without this avowal it is difficult to see how Britain's commitment to the defence of the Continent could ever have progressed beyond reluctant involvement. Here, in early spring 1950, one year after the signature of the Washington Treaty, some six months after the creation of the West German state and at a time when the British were, supposedly, underlining their commitment to the defence of the Continent by promising to despatch reinforcements there in the event of war, senior military staffs can be found articulating their perception of the threat, and Britain's response, in very limited terms. Germany, the Netherlands, Belgium and France were not to be defended out of some sense of fraternal good-will, but because their security was necessary for Britain's defence. It is this mind-set which explains the relatively low priority given to the Army in British military planning in these final pre-Korean War months. Britain's commitment to the defence of Europe was still very much a hollow one, best seen as an attempt to establish – by persuasion rather than force – a defensive glacis against aggression from the East.[40] Britain's main defence effort was still to be in the form of air power.

The second milestone to emerge from the COS discussion in mid-March 1950[41] concerned Britain's impression of the Soviet threat to the Middle East. During discussion it was suggested that some reference be made to 'the fact that it was now appreciated that the full threat in the Middle East would not develop until about six months after the war had begun': the COS accepted the recommendation. This comment is the first indication in the 1950 papers that a fundamental change of view was taking place in respect of the priority accorded to the Middle East in British defence planning, and this should be examined in some detail.

The COS returned to the subject four days later and the minutes of this latter discussion paint a clearer picture of what had taken place. A confidential annex records discussion of Plan 'Galloper', the combined Anglo-American plan for evacuation and return in the event of war in the short-term, up to July 1951.[42] The JPS paper was largely an adaptation of the previous short-term British plan ('Speedway'), in the light of recent events. One such event, to which the JPS referred, was the 'radical change in the assessment of the threat to the Middle East'. The source of this reassessment is given as JIC (49) 80 (Final). Later in the paper a more informative reference to this seminal JIC document can be found:

Probable Soviet operations at the outbreak of war were estimated by the [JIC] to be:

(a) Simultaneously

   (i) A campaign against Western Europe including Italy.

      (ii)   An aerial bombardment against the British Isles.

      (iii)   Campaigns against the Middle East, including Greece and Turkey.

      (iv)   Campaigns with limited objectives in the Far East.

      (v)   Attacks with limited objectives against Canada and the United States, including Alaska and the Aleutians.

      (vi)   A sea and air offensive against Allied sea communications.

      (vii)   Subversive activities and sabotage against Allied interests in all parts of the world.

      (b)   As soon as possible, after the occupation of the Channel Port areas, a full-scale sea and air offensive against the British Isles.

      (c)   As soon as practicable, campaigns against Scandinavia and possibly against the Iberian Peninsula.

> The [JIC] consider that the Soviet Union would have sufficient armed forces and resources *to undertake all the campaigns listed above and still retain an adequate reserve* [emphasis added].[43]

> An opportunist campaign by Chinese Communist forces, which may or may not be directed by the Soviet Union might be undertaken at any time.[44]

What was beginning to take place was a reduction of emphasis on the Middle East in British defence and foreign policy. Two reasons might be suggested for this swing. First, the assumption that Middle Eastern oil was a strategic necessity was now being questioned by military staffs in Britain. That JPS Plan 'Galloper' could contain a comment to the effect that 'the recapture of the Middle East oilfields, *should* this oil prove essential for the continuation of the Allied war effort, *may* be considered a *possible* operation' for the Allies in the period D+6 months to D+12 months (emphasis added),[45] suggests a far more flexible view of the strategic value of the Middle East than previously. So far, the strategic orthodoxy had been that the 'three pillars' – defence of the UK, protection of SLOC and maintaining a 'firm hold' in the Middle East – were all vital:

> These three pillars of our strategy must stand together. The collapse of any one of them will bring down the whole structure of Commonwealth Strategy.[46]

The second explanation, quite simply, is that the Americans had changed their strategy. By spring 1950, Britain's strategic planners had lost much of the self-confidence they had earlier displayed regarding Britain's strength and her place in the world, relative to the US. The planners thus came under mounting pressure to keep in line with American thinking. The same applied to the JIC: JIC (49) 80 (Final) – the main source of the change of assessment in Britain of the threat to the Middle East and to Britain generally – was to some extent a joint American, British and Canadian effort. The reluctance or inability of the Americans to commit troops to the defence of the Middle East, combined with (or, more likely, leading to) the British planners' own assessment that the threat to the Middle East would not materialise for six months, meant that a more relaxed view of the strategic importance of the Middle East could develop.

The early months of 1950, marked by a reassessment of the relative importance of the defence of the Middle East and Europe and by the need to keep pace with changes in American views, were clearly a turbulent time for British strategic thinking. But out of this turbulence a new, general strategy was developed, designed to replace the plan which had been drawn up in May 1947. The new strategic review process had begun in late 1949, with the goal to make an appraisal of the new global strategic situation, decide whether existing strategic thinking could meet any new demands being placed on it, and make new plans as necessary. The outcome, the last major strategic assessment before the outbreak of the Korean War, was a paper entitled 'Defence Policy and Global Strategy' (DO (50) 45), produced between May and June 1950.[47] Where the paper comments, albeit briefly, on the Soviet threat to Europe, Germany and Britain, it amounts to the final pre-Korean War threat assessment and is thus of special interest here.

The COS saw two types of threat, the first pertaining to the Cold War and the second to its successor – the 'shooting war'. During the Cold War there would be a need to guard against Fifth Column activities, the aim of which would be to 'rot resistance from within'.[48] As for the 'shooting war' in the Middle East, the COS openly accepted that 'the full [Soviet] threat is unlikely to develop against Egypt until considerably later than we had hitherto thought probable'. Egypt was seen to be the 'key strategic area of the Middle East',[49] just as the UK was 'the key to Western defence'.[50] The COS therefore expected that in any 'shooting war' in Europe, the UK would be singled out for special attention. Yet they appeared to take the analogy with the Soviet threat to Egypt and the Middle East even further when they commented they did not 'now regard a shooting war as inevitable or even likely'.[51] At first glance, this comment suggests that in the last months before Korea the COS had relaxed their assessment of the danger to Europe posed by the Soviet Union. Elsewhere, however, the paper shows that the sense of danger was in no sense reduced; 'from the purely military point of view, Russia could march to the Atlantic at any moment'.[52] But this was not a sufficient basis for a strategy which could serve in both peace and war, and it is in their awareness of this that the COS demonstrated the subtlety of their thinking. Far from being alarmist and simplistic, their analysis of the 'worst case' of massive Soviet aggression in Europe brought them instead to the conclusion that Allied strategy was as much a matter of political cohesion and steadfastness within the West as military plans and preparation. Soviet aggression was not likely, but it was certainly possible, and this possibility had to be kept firmly in mind if the political and ideological 'defences' of the West were to be maintained. These were, in fact, the first line of the West's defence, and if weaknesses developed the Soviet Union would be sure to exploit them, thus making war more likely.

As late as spring 1950, British strategic thinking continued to focus upon the air threat and the air response – a real territorial strategy was still very much out of favour. And the last major threat assessment before June 1950 only served to confirm the value of the air power deterrent. The continental school, such as it was, might have been expected to emphasise the possibility of war with the Soviet Union in the short and medium term and the need to support plans for the defence of Western Europe. But this would have been to fly in the face of strategic orthodoxy and would in any case, given the lack of resources, have been unsupportable. The much-vaunted reinforcement decision of March 1950, whereby BAOR would be reinforced with two divisions, was carefully hedged: as far as the Army was concerned, the reinforcement would only take place in the event of war, and would merely involve bringing reservists and territorials up to standard as fast as possible. It was not decided to reinforce BAOR with regular troops. In short, the March 1950 decision did little damage to the prevailing doctrine of 'limited liability', articulated by Liddell Hart in the 1930s, where commitment of forces to Europe was concerned:

> Britain should not again raise a large conscript army, put it on the Continent and use it in an offensive strategy in pursuit of total victory. Salvation lay rather in returning to traditional ways. Britain should leave major land-fighting to her Continental allies and concentrate on building up strength by sea and air. ... it would probably be best to commit no land force to the Continent at all, and make it quite clear that in any future conflict Britain's liability should be strictly limited.[53]

Resource allocation

Even before the end of the war in Germany, the Chiefs of Staff had largely accepted the need for rapid and large-scale retrenchment in Britain's armed forces – particularly the Army. A Joint Planning Staff paper entitled 'Manpower One Year After the Defeat of Germany – Reduction of Service Requirements'[54] suggested reductions in the armed services of between 500,000 and 700,000, and saw occupation tasks in Europe as just one of four 'Strategic Requirements' which the planners expected to meet one year after the end of the war in Europe (the other three were security of the United Kingdom, the war against Japan and the security of imperial lines of communication, India and the Middle East). According to the JPS, the occupation of the British zone of Germany would require the following British forces:

- A 'Mobile Reserve' comprising:
    1 infantry division[55]
    1 armoured division
        Sub-total: 50,000

- Occupation forces comprising:
    7 occupational groups[56]
    5 armoured car regiments
        Sub-total: 141,100

- 'Non-effectives' (i.e. administrative staff):
        Sub-total: 14,400

- Civil affairs personnel:
        Sub-total: 4,000
        Total: 209,500.[57]

In both its size and its duration the occupation commitment appeared to the military staffs to provide scope for reductions. The object of the occupation of Germany was simply to ensure that the instrument of surrender was carried out, and to maintain law and order in the British zone and Berlin; short-term, non-operational goals.

In mid-August 1945, with the Japanese all but defeated, pressure began to mount for further reductions throughout the services. In September the COS attempted, with some indignation, to reassess their estimates.[58] As far as the occupation of Germany was concerned, the winter of 1945–6 was felt to be a critical period, and it was again acknowledged that air power could be a valuable means with which to 'enforce authority'. By this stage, the preferred military commitment to the occupation of Germany was of the following order:

1 armoured division
2 infantry divisions
1 armoured brigade group
7 occupational divisions
5 armoured car regiments

The COS remained under pressure to provide the leanest possible estimates, steadily reducing their forecasts in all categories.[59] Other occupation commitments, however, offered better opportunities for the Army: British troops to be deployed to India amounted to twenty-three infantry battalions (with four divisions, five brigades and ninety infantry battalions coming from

India), while the Middle East was to receive one armoured, two infantry and one airborne division, seventeen infantry battalions and three armoured car regiments, together with a substantial contribution from the Indian Army.

Plainly, in these first post-war months, Treasury and military were not deeply divided over the need for retrenchment, at least where the commitment to Germany was concerned. Nevertheless, a rift between the Treasury and the Chiefs of Staff was soon to develop. During 1946 the COS met constant demands, first for predictions as to the size and cost of forces required at given dates in the future, and then for reductions to be made in those predictions.[60] But COS estimates were, predictably, always too high: in December 1946 Minister of Defence Alexander noted that COS estimates amounted to a figure which equated to the entire revenue from income tax for 1946.[61] By this stage, the COS were beginning to realise that their best hope was to present a united front to the government. Yet while they accepted that the problem was one of too few resources with too many commitments, they could not accept that it was they who should decide which commitments could or should be liquidated. Instead, they sought to pass unpleasant decisions back to the government. If commitments were to be reduced, this would be against their strategic advice and the government would either have to 'guarantee' a period of no threat to British interests or accept the risk.[62] Nevertheless, under mounting budgetary pressure inter-service loyalty came under ever more strain and the Chiefs of Staff increasingly defined their responsibilities and requirements in terms of their own service. In the face of government depredations, their loyalty to each other was not unlimited: on one notable occasion in December 1946, Montgomery's absence in Greece led to his Naval and Air Force colleagues suggesting that the bulk of the cuts could go on the Army rather than on their own two services.[63]

In June 1947, the COS examined projections of the strength of the forces for 31 March 1948, in response to Alexander's request for guidance as to the tasks facing the forces in the remainder of FY 1947–8 and for indications of 'the risks involved should the run-down of the Forces be increased'.[64] As in the past, the COS decided to pass the matter on to the service ministries and central staff. But what also emerges from this meeting is that long-term retrenchment was no longer sufficient: current commitments were now under scrutiny. This had obvious implications for the Army. The size of the Army on 31 December 1946 had been 896,710 (*c.* 63 per cent of the total armed forces).[65] The War Office now began to argue that while a reduction to 590,000 (*c.* 54 per cent of 1,087,000) could be achieved in the time required, this would be at the expense of the technical efficiency of the Army and would reduce the core of experienced soldiers. The figure of 590,000 was offered as an absolute minimum, with the War Office arguing that a figure of 672,000 would be a more appropriate estimate for 31 March 1948. By mid-summer 1947, War Office opinion had hardened. The Army now argued that the figure of 590,000 could only be achieved by making 'very drastic cuts' and by accepting a number of assumptions, *inter alia* that 'there will be no deterioration in the situation in Palestine or Germany, and a reduction of approximately 40 per cent in BAOR between 1 October 1947 and 1 April 1948 can be accepted'.[66]

The pressure for force reductions was relentless, and the cracks between the services began to widen. During the summer, the COS discussed memoranda from the three service ministers on the size and composition of forces required for 1948–9 and the years immediately following. Tedder, for the RAF, felt that the War Office had not gone far enough, but Sir Frank Simpson, Vice-Chief of the Imperial General Staff, fought back. The planned reductions had been tailored to meet what were being assumed to be the roles of the Army: the maintenance of essential overseas garrisons in support of foreign policy; and the training of the National Service intake.[67] In consequence, said Simpson, 'The Army was no longer therefore designed in relation to readiness for war, the only contribution to which would be the Territorial Army.' Further reductions could only be contemplated if certain overseas commitments were eliminated. Alexander, frustrated by the absence of a clear and unified response, warned of 'an arbitrary cut of all three Services ... which would not be related to strategic considerations or the advisability of maintaining balanced forces'. 'Deploring' the lack of agreement among the COS, Alexander left the meeting. Montgomery's response to the deteriorating atmosphere was to condemn the 'imposition of percentage cuts' as a return to 'the old system of arbitrarily cutting each of the original service proposals'.[68] Montgomery also refused to accept that the JPS should be given the complex task of examining the size and shape of the armed forces as a whole, and making recommendations. It was as if Montgomery recognised that the Army could best be protected by maintaining the impasse with government.

Competition for resources – particularly between the Air Force and the Army – resumed in December 1947, during discussion of the likely role of the Army at the outbreak of war. Tedder considered that the Army's tasks 'could, in general, be limited to the defence of the United Kingdom and bases abroad, including anti-aircraft defence'. Simpson, however, dismissed Tedder's reliance on the deterrent capability of 375 atomic-capable bombers as not 'realistic' and recommended that 'The only deterrent ... to a future aggressor would be the knowledge that they would have to face the combined strength of the United Kingdom and the United States'. Simpson insisted that the threat to the UK and overseas bases would also include airborne invasion, and that to ensure the security of these areas a force of 9½ divisions (TA), a small regular army and an anti-aircraft defence command would all be needed.[69]

The deadlock within the COS, and between the COS and the Cabinet, had discernible effects upon planning for Germany, and upon perceptions of the role and structure of the Army's presence there. In mid-February 1947, Alexander had acknowledged the danger of a 'Nazi irredentist movement' and was anxious to know whether British forces would be able to meet such a threat. Should the Army be 'more operational', leaving policing matters to a regenerated German police force? Simpson's response to Alexander was telling:

> We had three field divisions in Germany. It was the policy to concen-
> trate these forces as much as possible to be prepared to deal with any
> trouble. Owing to accommodation difficulties, however, it had not been
> possible to concentrate our forces as much as desired. Further, our
> forces were taking part in the administration of Germany by helping the
> Control Commission in such matters as workshop facilities and feeding
> arrangements.[70]

This does not, of course, give the impression of an operationally organised
force deployed to meet a military threat of any description. But neither
should it be inferred that the Army was losing interest in the occupation
commitment to Germany. In May, while describing the Army's major
commitments under the new defence policy (DO (47) 44), Montgomery also
referred to 'other Army commitments':

> Although it will probably prove necessary to withdraw our forces from
> Germany on the outbreak of war. ... It is ... necessary to consider our
> control forces in Germany as additional to our immediate requirements
> elsewhere. We should hope that within the next ten years the forces
> required for Germany should not exceed 20,000 men of whom the
> majority would be infantry.[71]

The Army was ever more deeply concerned that retrenchment might be
allowed to go too far. At one point Sir Richard McCreery, Commander-in-
Chief BAOR, warned that BAOR was in the process of transition from a
strong, experienced and efficient field army to a large scale training organisation', and was in danger of becoming a 'military kindergarten'.[72]
McCreery considered that further run-down in Germany could lead to
internal security problems. During the summer, he offered a concise account
of the tasks then facing BAOR:

> the primary operational task of (BAOR) will remain as at present, to
> aid the civil power in the maintenance of law and order and enforcing
> Military Government orders in the British Zone of GERMANY. In the
> event of a peace treaty being signed, (BAOR) will assist in ensuring that
> the terms of such a treaty are carried out, and will enforce sanctions if
> necessary when this is NOT done.[73]

According to McCreery, the type of army units needed in BAOR were 'of
high mobility with the ability to search; great fire power is NOT required. It
follows that the most suitable units for the operational task are armoured
car regiments and infantry battalions'. The RAF would not be needed.
Given that the most suitable units to meet a Soviet military threat in
Germany would have been armoured divisions and brigades, with massive
air cover, it is worthy of note that in the view of the senior British combat

soldier in Germany, the tasks facing British forces were essentially internal security and of German, rather than Soviet, origin.[74]

With solidarity among the COS dwindling under budgetary pressure, Montgomery nevertheless persisted in his stonewalling tactic: passing back to the government difficult decisions about the size and shape of the Army, and its future strategic role. In May 1948 in a memorandum entitled 'Preparedness for War', Montgomery argued that the Cabinet should be asked to decide whether, in view of present circumstances, war was considered inevitable. If so, when would it occur and what resources would be allocated to the services to prepare for the conflict? Montgomery could no longer accept that war – be it 'accidental' or otherwise – would not break out before 1957:

> It is, therefore, unrealistic to work on the philosophy that there will be no war for a predetermined number of years. We are not ready for war and, as a result of economic and manpower stringency, can make no really adequate preparation. We should ask Ministers how long they are prepared to accept this state of unreadiness.[75]

Tedder responded by arguing that while the Navy and Air Force had produced their plans for meeting a crisis in 1957, the Army had so far failed to produce an equivalent, and had persistently concentrated on their 'peace-time commitments, i.e. the maintenance of occupation forces, the reorganisation and training of the Territorial Army and of the National Service intake'. Sir John Cunningham, First Sea Lord and Chief of the Naval Staff, suggested that, since the COS had agreed to provide the Defence Committee with a long-term plan by the autumn, the Army should work on an order of battle for 1957 designed to meet the agreed three pillars of strategy. Occupation forces, he felt, should be accounted for in a supplementary estimate. This was not acceptable to Montgomery: not only did he want to protect the Army from future cuts by making ministers politically liable for any more cuts, but he was also clear in his own mind that the maximum benefit for the Army should be extracted from whatever commitments – such as occupation – it did have, however peripheral and short-lived.

At a meeting in June, Montgomery developed another approach.[76] Warning loftily of the dangers of uncoordinated planning and 'acrimonious' inter-service wrangles, Montgomery proposed three principles for planning. First, the only way by which the allies would be able to knock out Russia in war was by 'the development of air power'. Second, to enable the first, the three pillars must be maintained. Finally, to secure an effective air base in the west it was essential that Russia be held as far east as possible, allowing bomber forces to operate from the UK and Western Europe.[77] Thus, while ostensibly bowing to the might of the air strategy, Montgomery ensured that the Army would be given the crucial role of securing air bases from which the air offensive could be developed: 'We must win the first battle first, if we are to be in a position fully to develop our air power in the decisive battle.' Sensing a rare opportunity,

Montgomery began to contradict his earlier position regarding the level of commitment of troops to Germany, talking instead of a BAOR of four divisions. Tedder, of course, resisted, and argued that the Army should join fully in the long-term planning process which, in the contemporary technological and economic climate, could only be damaging for the Army.

But then came the Berlin crisis, which was something of a gift for the Army. Judging the moment now to be right to enter into long-term planning, Montgomery prepared a paper entitled 'Role of the Army on the Outbreak of War and Plan for Long Term Development of the Army to fulfil its Role',[78] in which he suggested as many as six roles for the Army:

- The land battle in defence of the Western Union;
- Helping to secure Britain's position in the Middle East;
- Garrisons abroad for the defence of sea and air bases vital to the maintenance of SLOC;
- Civil defence in the UK;
- Air defence of the UK;
- The nucleus for expansion.

Montgomery now argued that Britain should deploy to Europe two army corps of two divisions and supporting arms each, in order to provide support to the Western Union, and he dismissed previous attempts at costing the size of the Army, either because they had been based on 'a very different role', and had not taken account of the Western Union, or because they had envisaged a 'no war' period and had not provided for preparation for war.[79] Ever the pragmatic opportunist, Montgomery saw how the Berlin crisis could best be exploited for the Army, but was also wary of promising too much or of being seen to undermine the air power strategic orthodoxy:

> Our forces in Germany, though they were not strong enough physically to prevent the Russians from sweeping across Europe, were a most important factor in stimulating the moral resistance of the Western Union Nations to Communism, and thus from preventing a war from breaking out.[80]

As the strategic debate moved into 1949, the Army took an increasingly robust attitude. During discussion of the report of the Harwood[81] interdepartmental working party, set up late in 1948,[82] the COS now accepted the Army's argument that the requirements of the Cold War and other peacetime commitments would involve larger land forces than might otherwise have been necessary. But Slim, having succeeded Montgomery as CIGS, wanted to go further, complaining that the Harwood Report had not made enough of an 'active cold war policy' which he felt could be an additional, positive deterrent to war. Force planning ought, he felt, to reflect the assumption that Britain could conduct 'an offensive cold war policy aimed at separating Russia from her satellites and disintegrating Russia from within'. A disagreement then developed between Slim and Tedder, each trying to show that 'fighting the cold war' was a vital task performed best by his own service. Tedder resented the assumption in the Harwood Report that the Cold War was primarily a task for land forces: the Berlin airlift, 'perhaps the most important single factor in the cold war', was a 'severe strain' for the RAF and so, by implication, it was the RAF and not the Army which should be assigned any additional funding. Although the gulf between Army and RAF was deepening, the COS nevertheless saw the value of presenting a united front to the government and described the Harwood report as 'courageous' and 'a tribute to the Joint Service spirit'. As far as the government was concerned, the COS had to be unequivocal: without a larger budget, commitments would have to be cut or re-equipment would have to be limited. But the underlying differences between the services proved irreconcilable: as each service analysed the Harwood Report in detail, so the penalties of accepting the report became apparent, and in the end the COS were unable to recommend the Harwood Report to the Defence Committee as the basis for defence planning and budgeting.

The undermining of the Harwood Report prompted the establishment of a new working party: the Inter-departmental Committee on Defence Estimates, chaired by Sir Harold Parker,[83] charged with drawing up a provisional defence budget for April 1950 to March 1953. At the request of the committee, each service produced its own estimates for the three years in question: but as ever, the overall totals considerably exceeded the figures permitted. Parker asked for strategic guidance as to where reductions could be made, and the COS duly met to discuss the problem.[84] Lamely, the COS were unable to offer Parker's committee much help or advice, even though they knew the Defence Committee was waiting to make plans for the next three years. As so often, the COS tried to judge their response carefully, anxious to appear sufficiently constructive and thus avoid being labelled irresponsible or unrealistic and having arbitrary cuts imposed upon them.

The response of the War Office to the Parker Committee enquiry is worth examining in some detail, for the light it throws upon the Army's conception of its current and future roles.[85] 'Victory in the Cold War', the War Office paper claimed, was the 'first essential'. The Cold War was expected to continue for several years and should be pursued even at the risk of delaying preparation for a premeditated war. As a concession to the possibility of war breaking out, the War Office was prepared to lay no more than the foundations for rearmament for 1957 over the next three years but felt that, if an unpremeditated war did break out, the only course of action was to do the best possible with the available resources. To wage the Cold War, an Army of around 355,000 would require equipment and resources to resist Communist pressure, maintain order, provide moral support, arms and other material support and fulfil treaty obligations. These resources would not require huge expenditure, suggesting that the main objective was a simple one: a large Army. Thus, the War Office listed the orthodox strategic requirements at the outbreak of war, taking issue with none of them, but claimed that the waging of the Cold War would also require:

> A clear demonstration that we will resist armed aggression, have plans ready for that contingency, and retain a firm hold on those bases from which we would launch our air offensive.

By this argument the War Office's plan for the Army – involving a large uniformed strength but little in the way of costly equipment or other resources – was tied firmly to concepts which might otherwise have undermined the Army's *raison d'être*. Later in 1949 Slim made the Army's position in the strategic debate clearer than ever before. For the Army, the most fertile ground was plainly not planning for an unpremeditated war in 1957 or thereabouts – all signs pointed to the RAF as the most appropriate means with which to wage such a war. Rather, the confusing and strategically novel Cold War was emerging as the Army's safest bet. Slim was, accordingly, 'obliged to neglect readiness for an unpremeditated war and was concentrating on the cold war'.[86] At the end of September, in a COS report to Alexander entitled 'The Size and Shape of the Armed Forces over the Next Three Years', the Army's position was made clearer still:

> The greater part of [the Army's] units are already deployed overseas to meet Cold War commitments. There is little prospect of these commitments being reduced – the tendency is for them to increase – and if they are to continue to be met, there can be no reduction in the major units of the regular Army nor, unless regular recruiting greatly improves, in the length of National Service. Unfortunately, the demands of the Cold War are, as far as the Army is concerned, largely antagonistic to preparation for a Major War. The Army, on any possible Budget, cannot effectively bear the burden of the Cold War and at the same time prepare fully for a Major War. One must have a preference; this should be the Cold War.[87]

In early 1950 the familiar pattern seemed about to repeat itself, with Shinwell, the new Minister of Defence, calling for reductions and the COS threatening to pass the whole problem back as a 'political' matter. At this stage, short-term planning for 'hot' military operations in Germany was not seriously considered; a deficiency which would have to be accepted. This suited not only both Shinwell and the Treasury, for obvious budgetary reasons, but also for very different reasons both the Royal Navy and RAF, who saw in long-term planning an opportunity to diminish the effect of budgetary pressure on strategic thinking. The Army, as we have seen, was in any case not willing to enter into tri-service, long-term planning for military operations. But then, with the invasion of Korea, the agenda began to reflect much more favourably the Army's position, as the need for short-term planning for military operations in Europe was perceived more keenly. The shock of the Korean War provided the impetus towards a more military ratio-nale for the involvement in Germany. The Army grasped with alacrity the opportunity to meet the other services on more equal strategic and resource terms. But the result – the March 1950 reinforcement decision – was care-fully hedged by the Army's reluctance to be drawn into planning and preparing exclusively for another major campaign in Europe.

Summary

For the first two years after the defeat of Germany, Germany, and Britain's occupational commitment there, were consistently at the bottom of the COS list of priorities. The commitment to the defence of the Middle East and the sea lines of communication, the need to organise and deploy an effective Imperial Strategic Reserve, and the fascination and faith in air power and the inchoate doctrine of deterrence through massive atomic retaliation, all took pride of place in British military thinking during this early post-war period. In this context, not only the type of commitment which the Army might make, but also the Army's very existence were subject to some awkward and uncomfortable scrutiny. By May 1947 the British military view of Europe was, in every sense, minimalist. There was no emergency planning for war in Europe in the short-term with the Americans or even nationally. Britain's military presence and interest in Germany during these years was an occupa-tional and policing matter and not 'operational' in the sense of erecting a defence posture against the Russians or any other potential enemy.

Following the uncertainty and minimalism of the immediate post-war period, from the end of 1947 military thinking about Germany became more certain, albeit in a negative sense; there was now even less willingness to entertain a continental strategy. The basic budgetary, technological and doctrinal prejudices against military involvement on the Continent went largely unchallenged and were even given something of a fillip during inter-ventions by Attlee at the end of 1947.[88] The Future Defence Policy paper (DO (47) 44), which provided the framework for strategic thinking during 1947 and beyond, implied further movement away from military involve-ment in Germany. And from early 1948 Britain was involved in new planning discussions – most notably those with the Americans – which, again, postulated movement away from Germany both conceptually and physically. The air/maritime strategic orthodoxy remained potent into 1948. Time and again, faith in air power and the atomic strike was confirmed, and the three-pillared defence policy underlined. This orthodoxy was rooted not only in doctrinal and technological preferences but also in severe limitations on resources. Britain's commitment to Europe began and ended with the occupation of Germany; a commitment which was certain to be run down.

From early in 1948 Britain's military planners were pulled in opposing directions. In one direction lay Anglo-American planning for an emergency withdrawal from Europe: the contingency most compatible with current British military thinking.[89] In the other direction, however, lay the integration and reconstruction of Western Europe, encouragement of which necessitated some form of defence commitment. Strategic thinking then deteriorated into a series of attempts to find a compromise and to divert European interest into less difficult areas: long-term force planning, for example. The image of Montgomery championing the continental strategy in the face of the scepticism of his colleagues and the government is not the most complete representation of what occurred, largely because the air/maritime orthodoxy was unassailable. Montgomery's motives and actions were more subtle than might be supposed. He never set out to challenge the strategic orthodoxy, but sought instead to establish a stronger position for the Army within that orthodoxy and to manipulate the planning and resource allocation processes to the Army's benefit. Thus, in 1947, he emphasised the strain imposed upon the Army of the National Service training programme, of the need to defend overseas commitments (especially those on the SLOC), and of the need to meet a possible invasion. But more significant was his attempt in 1947 to arrest the long-term planning process altogether, ostensibly for the sake of the nation's economy but actually because the Army could only fare badly in such a process.

The development of the Western Union and the requirement for some form of military commitment by Britain ought to have provided Montgomery with an opportunity to strengthen the Army's position. Instead, he advocated no more than a token commitment of land forces which he accepted would be limited, ill equipped and late in arriving. Equally, the effect of the Berlin crisis on Montgomery's thinking was not straightforward. Montgomery emphatically did not see the defence of Berlin as providing an immediate, key role for land forces. Rather, he was concerned that involvement in Berlin – a city he had 'written off' – would detract from the more important and realisable goal of withdrawal to, and defence on, the Rhine. In response to the crisis, Montgomery reiterated his calls for a British commitment to a strategy of defence for Europe. But the commitment he sought, once again, neither challenged the orthodoxy in Britain nor laid any particular stress on the value of land forces. Having first been concerned that the crisis would overshadow long-term planning, Montgomery finally saw in the Berlin crisis the long-awaited rationale for the abandonment of the long-term planning process altogether, and the concentration instead upon more favourable peace-time considerations and deployments. Even then, his argument fell noticeably short of challenging the strategic certainties of the day.

The British Army's interest in Europe during 1947 and 1948 might best be described as an exercise in public relations on a continental scale. There is little evidence of steady progress towards a continental strategy during these years, not least because, after Montgomery's departure for the Western Union, the War Office began to focus more on the Middle East than on

Europe, and on Anglo-American rather than Western Union planning discussions. Slim's response to Montgomery's pleading on behalf of the Western Union was merely the vague promise that occupation troops in Germany would be made more battleworthy. Above all, it is clear that 1948 ended with a wide discrepancy between the developing political commitment to Europe and its defence, and the strategic thinking of Britain's military, particularly the Army.

By late July 1949 the COS had agreed grudgingly to make just one division of the occupation troops already in Germany battleworthy by the end of the year. The COS and the Joint Planning Staff had become steadily more aware of the broader political dimensions of the situation in Germany and had come to accept that there was a political or Cold War requirement to prevent the collapse of the Western Union. A reinforcement promise might contribute to this end. This did not, however, mean that the COS and JPS had in any way altered the underlying assumptions of strategic thinking in Britain. There were two such assumptions. The first was that the most appropriate way to deter and then fight and defeat the Russians was with a massive strategic air offensive using atomic weapons. The second was that the Middle East was vital to the survival in war of Britain and the Commonwealth. Defence of Europe was not entirely abandoned by this rationale: the strategic air offensive could be launched most effectively from the UK, and so it was essential that the UK be defended. If a defensive perimeter could be held in Europe, so much the better. However, both the British and Americans considered the chances of holding the Russians in Europe to be so slim that the likely loss of Europe had been accepted. Thus, from the British perspective, defence of Europe was qualified in two ways: first, it was not an objective in its own right, but derived from the need to maintain the UK strategic air base; second, it was a task to which Britain would be both unable and unwilling to allocate any military resources other than those which happened to be in Germany at the time, i.e. the under-equipped and badly trained occupation forces. Out of this mix of political and military perspectives grew a series of suggestions regarding the formulation of a reinforcement promise. In each case, the aim was to stimulate European morale while constructing the offer in such a way that it would probably never come to anything.

The basic assumptions driving British strategic thinking were maintained up to March 1950. Thus, it was decided in advance that a review of defence policy and global strategy should not move in the direction of a continental strategy, and even the War Office argued vigorously against such a development. What characterised this final, pre-Korean War debate, however, was the challenge made to these assumptions by a new American 'Center Strategy' launched early in August. The defence of the UK as a base for the strategic air offensive remained the priority, but while the British gave second place to the defence of the Middle East, the Americans now stressed the need to defend Western Europe. The new American plan was at first just an extension of the existing plan to abandon Europe in the event of war: the novelty was merely that a bridgehead or lodgement would be held from which a major land counter-offensive would eventually be launched. In terms of the likelihood of a successful defence of Europe against the Russians, the Americans remained just as pessimistic as the British. The problem for the British, quite apart from the fact that the Americans had diverted a large proportion of their strategic air strength to tactical tasks in Europe, was that the reallocation of forces to the defence and expansion of the European lodgement would mean that the British would be left alone in the Middle East and incapable of mounting an effective defence. Eventually, the Americans were persuaded to adopt a more flexible posture and agreed to reinforce a defensive position on the Rhine in war, provided there were signs that it was being held.

The main British goal in this was to deflect anticipated French demands for a more substantial British guarantee, ensuring that the Middle East would not be completely abandoned. Thus, when the British military seemed by the end of 1949 to have accepted a more direct military interest in European defence, this was more apparent than real, and in any case reflected external pressure rather than a change of heart on the part of the British. The result of this enforced reappraisal of strategic goals was the reinforcement decision of March 1950. Significantly, the troops being offered would be Territorial Army formations, requiring three to six months to prepare for war, and would be of questionable value even if they arrived in time to make a contribution. Furthermore, these divisions would be allocated only very limited tactical air support: an important indicator of British interest and sincerity, since without such support the life expectancy of Army divisions fighting in Europe would not be long. Quite apart from the substance of the reinforcement decision, it is important to note not only that the British been pushed in this direction by their American allies, but also that, by attaching so many limitations and qualifications to the reinforcement promise, the British clearly still viewed European defence as no more than defence of Britain by proxy.

Conclusion

Between the end of the war in Europe in 1945 and the start of the Korean War in 1950, rather than identify and learn the key lessons of its intense and hard-fought campaign in Europe, the priority for the British Army was to find ways to survive in the battle for credibility and cash in Whitehall. The Army shifted its emphasis and approach as circumstances changed. In the first, post-war months, with cuts in manpower inevitable, the Army's priority was to ensure that cuts would be manageable and paced steadily. When it appeared that cuts in Army strength might go too far, the Army argued for a united front among the COS, in order to stonewall the government and place the onus for risky strategic decisions on the Minister of

Defence and his Cabinet colleagues. Gradually, the Army realised the need for a more constructive approach, and saw that the much-denigrated occupation commitment was actually an important asset: not only was it a job the other two services could not perform, but it was one with strategic significance, insofar as a central danger was perceived to be the collapse of order and control in Germany and a consequent takeover by communists. All along, the Army recognised that the air-maritime strategic orthodoxy could not be challenged. But it could be exploited, by finding niches within the strategy (the defence of the UK and areas of Europe in order to guarantee bases from which to launch strategic air attacks) which only the Army could fill. And as far as planning was concerned, the Army sought to avoid it wherever possible: responsible involvement in long-term strategic planning and planning for re-equipment would only expose the Army to the conviction (which the Army shared) that there was little for the Army to do strategically in the long-term. All along, the main aim for the War Office was to keep the Army as large as possible, even if that meant being badly equipped, in the hope that more propitious circumstances might be around the corner.

Although, in the early months of 1950, the British military were beginning to appreciate both the political and the military dimensions of European defence, their interest in the latter was at best only indirect. The movement was clearly in the direction of a more European strategy but, as the Korean War broke out, there was still some distance to go. It is difficult to find, in the minds of Britain's military chiefs – including, most remarkably, the professional heads of the Army – much firm evidence of a reversal of Britain's traditional, maritime strategy – with its modern, air power overlay – during these pre-Korean War years. The British Chiefs of Staff were not even beginning to think, seriously, about a continental strategy for Britain. The irony in all this is that just as Liddell Hart was becoming less dogmatic regarding the territorial defence of Europe and a serious British contribution to it, so the Chiefs of Staff, including the CIGS, were struggling to sustain a view of British strategy which corresponded to Liddell Hart's pre-war notion of 'limited liability'. Kennedy has argued that the Second World War broke the 'myth of the efficacy of the 'British way of warfare' against a power which straddled half a continent'.[90] But this shift was not instantaneous; until at least June 1950 the 'myth' of limited liability was all too real. During the first post-war half-decade, the Army did not have the luxury of reviewing its recent performance and deciding how to shape its evolving Cold War force posture. Instead, it was fighting for survival in Whitehall and in the process was preoccupied with learning lessons of a new, more political kind.

Notes

1 J. Kiszely, 'The British Army and Approaches to Warfare since 1945', *Journal of Strategic Studies* (Volume 19, December 1996), p. 183.

2   Kiszely, 'The British Army', p. 183.

3   Kiszely, 'The British Army', p. 184.

4   See, for example, British Army of the Rhine, 'Battlefield Tour. First Day: 8 Corps Operations East of Caen, 18–21 July 1944 (Operation Goodwood)', G(Trg) HQ BAOR, June 1947.

5   This chapter draws on my *British Military Planning for the Defence of Germany, 1945–50*, London: Macmillan, 1996.

6   COS (45) 97th Meeting, 13 April 1945, CAB 79/31, National Archive, Kew.

7   For wartime threat assessments and planning for post-war strategy, and the division between FO and COS, see J. Lewis, *Changing Direction: British Military Planning for Postwar Strategic Defence, 1942–1947*, London: Sherwood Press, 1988, p. 131; A. Gorst, 'British Military Planning for Postwar Defence, 1943–45', in A. Deighton (ed.) *Britain and the First Cold War*, London: Macmillan, 1990, *passim*; G. Ross (ed.) *The Foreign Office and the Kremlin: British Documents on Anglo-Soviet Relations, 1941–1945*, Cambridge University Press, 1984, document no. 27; M. Kitchen, *British Policy Towards the Soviet Union During the Second World War*, London: Macmillan, 1986, *passim*, and especially p. 198; H. Thomas, *Armed Truce: The Beginnings of the Cold War, 1945–46*, London: Sceptre, 1988, p. 309.

8   COS (45) 11th Meeting, 10 January 1945, CAB 79/28.

9   COS (45) 252nd Meeting, 16 October 1945, CAB 79/40.

10 COS (46) 24th Meeting, 13 February 1946, CAB 79/44.

11 COS (46) 52nd Meeting, 1 April 1946, CAB 79/46.

12 COS (46) 128th, 131st, 135th Meetings, 22 August, 27 August, 3 September 1946, CAB 79/51.

13 COS (46) 54th Meeting, 5 April 1946, CAB 79/47.

14 DO (47) 44, Future Defence Policy, 22 May 1947: 'Should a resurgent Germany again become a menace, it would be possible to adjust our Defence Policy, if we have meanwhile prepared against a presently greater threat.' A copy of this document can be found in Lewis, *Changing Direction*, Appendix 7.

15 COS (47) 1st Meeting, 1 January 1947, DEFE 4/1. See also A. Deighton, *The Impossible Peace: Britain, the Division of Germany and the Origins of the Cold War*, Oxford: Clarendon Press, 1990, p. 140.

16 For discussions of the 'militarisation' of Britain's policy of containment, see J. Lider, *British Military Thought After World War II*, Aldershot: Gower, 1985, pp. 62ff. Lider identifies the notion of 'passive' containment and suggests that the COS began to see a military aspect to containment in 1947.

17 COS (47) 1st Meeting, 1 January 1947, DEFE 4/1.

18 COS (46) 54th Meeting, 5 April 1946, CAB 79/47.

19 See, for example, COS (46) 74th Meeting, 10 May 1946, CAB 79/48. The COS expected that, by late 1947, 'we should have largely reduced our occupation forces for Germany'.

20 COS (46) 102nd Meeting, 3 July 1946, CAB 79/50.

21 DO (47) 44, Future Defence Policy, 22 May 1947. Annex; National Policy, paragraph 3 (g). See Lewis, *Changing Direction*, p. 386.

22 See also COS (47) 34th Meeting, 28 February 1947, DEFE 4/2.

23 COS (47) 34th Meeting, 28 February 1947, DEFE 4/12. The JPS paper under discussion was JP (47) 4 (Final), Military Control of Germany, 26 February 1947. 'Phase 2' referred to the point, after the signature of the Peace Treaty, when a German government would be fully responsible and Allied forces would be reduced from tasks of 'occupation' to those of 'control'.

24 COS (47) 152nd Meeting, 8 December 1947 and 156th Meeting, 13 December 1947, DEFE 4/9.

25 COS (48) 30 (O), 7 February 1948, DEFE 5/10.

26 COS (48) 49 (O), 5 March 1948, DEFE 5/10.

27 Bevin to Alexander, C.3314/3314/G dated 27 April 1948. Copied at COS (48) 101 (O), 1 May 1948, DEFE 5/11.

28 Military Governor and Commander-in-Chief of British Forces in Germany.

29 Robertson to Bevin, MG/4103/C-in-C, 12 June 1948. COS (48) 142 (O) 25 June 1948, DEFE 5/11.

30 COS (48) 91st Meeting, 2 July 1948, DEFE 4/14.

31 COS (48) 164th Meeting, 17 November 1948, DEFE 4/17. Discussion of JIC (48) 104 (Final), a report on Soviet intentions and capabilities in 1949 and in 1956/57, and JIC (48) 100 (Final), an agreed Anglo-American intelligence appreciation of Soviet intentions and capabilities for the same periods. The latter document was approved as a background paper for further planning and intelligence studies.

32 The estimate of Soviet capabilities given in JP (48) 164 was rather more serious than that which the COS had accepted in COS (48) 30 (see footnote 74 above). By this latest count, the Russians would have some eighty-three divisions on the East bank of the Rhine by D+21 days.

33 This document was eventually to become the 'main assumption' behind the work of the Harwood Working Party. See COS (48) 176th Meeting, 9 December 1948, DEFE 4/18.

34 COS (49) 10th Meeting, 19 January 1949, DEFE 4/19.

35 COS (49) 14th Meeting, 28 January 1949; discussing JP (48) 117 (Final), DEFE 4/19.

36 JP (49) 3 (Final), discussed at COS (49) 55th Meeting, 13 April 1949, DEFE 4/21.

37 COS (50) 9th Meeting, 13 January 1950, DEFE 4/28.

38 JP (50) 22 Final, 10 March 1950, DEFE 4/29 and DEFE 6/12.

39 Ibid., para. 8. This conclusion appears to have been accepted in late 1949, as the following quotation from Minute 11 of COS (48) 188th Meeting, 21 December 1949, DEFE 4/27 indicates: 'there was now an increasing realisation that the security of the United Kingdom was very closely bound up with the defence of Europe.'

40 The values which the British military considered vulnerable to Soviet forces were western forces and strategic assets rather than the territory of Britain's continental allies and neighbours: JP (48) 131 (Final), Plan Speedway – Report by the [JPS], 18 November 1948, DEFE 6/7; JP (48) 140 (Final), Digest of Plan Speedway – Report by the [JPS], 8 December 1948, DEFE 6/7.

41 COS (50) 39th Meeting, 13 March 1950.

42 JP (49) 134 (Final), 1 March 1950, DEFE 4/29.

43 See COS (50) 100, Plan Galloper – Note by the Secretary, 28 March 1950, DEFE 5/20.

44 This comment is quoted largely because it shows that, months in advance of the outbreak of the Korean War, Britain's intelligence assessors were alive to the possibility of Chinese and/or Soviet adventurism, and gives weight to the thesis advanced by B. Heuser that Western governments anticipated problems in Korea; 'NSC-68 and the Soviet threat: a New Perspective on Western Threat Perception and Policy-Making', *Review of International Studies* (17/1, 1991), p. 25. For a contrasting interpretation of NSC-68 see M. Cox's response to Heuser; 'Western intelligence, the Soviet threat and NSC-68: a reply to Beatrice Heuser', *Review of International Studies* (18/1, 1992).

45 JP (49) 134 (Final), 1 March 1950, DEFE 4/29, para. 103(b).

46 DO (47) 44, Future Defence Policy, May 1947, para. 36. Quoted in Lewis, *Changing Direction*, pp. 370ff. A slightly less vehement expression of the same sentiment can be found in COS (47) 227 (O), *Review of World Strategic Situation*, 17 November 1947, DEFE 5/6. This paper is not quite so adamant that all three 'pillars' must all stand or fall together, but does contain a telling reference to the

strategic importance of Middle East oil: 'The importance to us of present and potential oil supplies in the Middle East is greater than ever. In the foreseeable future, unless measures are taken to change the situation, our dependence on Middle East oil seems likely to increase rather than decrease.' (para. 35).

47 The first and main version of the report is COS (50) 139, *Defence Policy and Global Strategy Report*, dated 1 May 1950. A slightly amended version of the report, aimed at an American, Australian, Canadian and New Zealand reader-ship, was circulated as DO (50) 45, dated 7 June 1950. It is this version of the report to which R. Ovendale refers in *The English-speaking Alliance: Britain, the United States, the Dominions and the Cold War*, London: George Allen & Unwin, 1985, pp. 122ff, and which R. J. Aldrich and J. Zametica described as repre-senting a 'shift towards continental defence' by Britain – 'The Rise and Decline of a Strategic Concept', in R. J. Aldrich (ed.) *British Intelligence, Strategy and the Cold War, 1945–51*, London: Routledge, 1992, p. 264. DO (50) 45, is published in its entirety in H. J. Yasamee and K. A. Hamilton (eds) *Documents on British Policy Overseas. Series II, Volume IV: Korea, June 1950–April 1951*, London: HMSO, 1991, appendix I, p. 411.

48 Ibid., p. 413; DO (50) 45, para. 11.

49 Ibid., pp. 422–3; DO (50) 45, paras 31, 33.

50 Ibid., p. 415; DO (50) 45, para. 13(b).

51 Ibid., p. 417; DO (50) 45, para. 17.

52 Ibid., p. 414; DO (50) 45, para. 13(a).

53 M. Howard, 'Liddell Hart', in M. Howard, *The Causes of Wars*, London: Temple Smith, 1983, p. 203; describing Hart's analysis of British strategy in the 1930s.

54 JP(44) 226 (Final), Manpower One Year After the Defeat of Germany – Reduction of Service Requirements, 3 January 1945, annexed to the minutes of COS (45) 12th Meeting, 11 January 1945, CAB 79/28.

55 The size of divisions in the mobile reserve may have been anomalous; calcula-tions made elsewhere in the paper are based upon the following strengths: armoured or infantry division – 40,000; airborne division – 20,000; armoured or infantry brigade – 13,000; independent battalion/regiment – 1,200.

56 Churchill was not content with the term 'Occupational Group' (proposed strength of 19,300) and later ordered it to be changed to 'Occupational Division'.

57 A further 80,000 troops would be provided for the British zone of occupation by France and Canada.

58 COS (45) 215th Meeting, 5 September 1945, CAB 79/38.

59 COS (45) 287th Meeting, 21 December 1945, CAB 79/42.

60 COS (46) 2nd, 7th, 12th and 15th Meetings, 4–29 January 1946, CAB 79/43–44.

61 COS (46) 176th Meeting, 3 December 1946, CAB 79/54.

62 COS (46) 23rd and 24th Meetings, 11 and 13 February 1946, CAB 79/44.

63 COS (46) 176th Meeting, 3 December 1946, CAB 79/54.

64 COS (47) 79th Meeting, 25 June 1947, DEFE 4/5.

65 COS (47) 39, Manpower – Strength of the Armed Forces at 31st March 1948, 19 March 1947, DEFE 5/1.

66 COS (47) 91st Meeting, 21 July 1947, DEFE 4/5.

67 These assumptions are mentioned in COS (47) 179 (O), Revised War Office Estimate, 29 August 1947, DEFE 5/5.

68 COS (47) 185 (O), Future Defence Policy; Size and Shape of the Armed Forces, Note by CIGS, 1 September 1947, DEFE 5/5.

69 COS (47) 152nd Meeting, 8 December 1947, DEFE 4/9.

70 COS (47) 24th Meeting, 14 February 1947, DEFE 4/1.

71 COS (47) 106 (O), The Army at the Outbreak of War, Memorandum by CIGS, 19 May 1947, DEFE 5/4.

72 McCreery to Sholto Douglas, BAOR/50854/9/AC, 11 June 1947. Quoted in COS (47) 136 (O), Internal Security Situation in the British Zone of Germany, 28 June 1947, DEFE 5/5.

73 COS (47) 159 (O), Germany – Internal Situation in the British Zone, 8 August 1947, (copy of a letter from Sholto Douglas to COS dated 7 August 1947), DEFE 5/5. McCreery's paper, The Future Organisation of BAOR, is dated 31 July 1947.

74 For something of the flavour of British Army life in Germany in the late 1940s, see R. Bainton, *The Long Patrol: The British in Germany Since 1945*, London: Mainstream Publishing, 2003.

75 COS (48) 63rd Meeting, 10 May 1948, DEFE 4/13. For Montgomery's memorandum, Preparedness for War, see COS (48) 104 (O), 6 May 1948, DEFE 5/11.

76 COS (48) 82nd Meeting, 16 June 1948, DEFE 4/13.

77 COS (48) 131 (O), Size and Shape of the Armed Forces; Inter-Service Preparation for War – Memorandum by CIGS, 11 June 1948, DEFE 5/11.

78 COS (48) 172 (O), Role of the Army on the Outbreak of War and Plan for Long Term Development of the Army to fulfil its Role – Note by CIGS, 9 August 1948, DEFE 5/12.

79 COS (48) 147th Meeting, 13 October 1948, DEFE 4/16.

80 COS (48) 174th Meeting, 7 December 1948, DEFE 4/18.

81 E.G. Harwood had been Civilian Director of the Imperial Defence College.

82 COS (49) 48th Meeting, 28 March 1948, DEFE 4/20.

83 Permanent Under-Secretary at the Ministry of Defence.

84 COS (49) 110th Meeting, 28 July 1949, DEFE 4/23.

85 COS (49) 247, The Future Shape and Size of the Army, 21 July 1949, DEFE 5/15.

86 COS (49) 132nd Meeting, 9 September 1949, DEFE 4/24.

87 COS (49) 313 (Final), The Size and Shape of the Armed Forces over the next Three Years – Report by COS, 27 September 1949, DEFE 5/16.

88 Attlee: 'As it was not envisaged we should send an army to hold off the enemy in Europe and since there was little risk of this country being invaded the necessity for maintaining a substantial army equipped with modern weapons and equipment appeared arguable.' (quoted in the minutes of COS (47) 158th Meeting, 16 December 1947, DEFE 4/9)

89 This was, however, a prospect that did not sit easily with the French government, whose commitment and contribution to European defence was increasingly being understood to be vital: M. Cresswell, ' "With a Little Help From our Friends": How France Secured an Anglo-American Continental Commitment, 1945–54', *Cold War History* (3/1, October 2002), p. 7.

90 P. Kennedy, *Strategy and Diplomacy 1870–1945*, London: Fontana, 1984, p. 80.

4 Lost and found in the jungle

The Indian and British Army jungle
warfare doctrines for Burma,
1943–5, and the Malayan
Emergency, 1948–60

Daniel Marston

In preparing this chapter, I have attempted to answer one of the principal
questions which is the focus of this book: did fighting in two world wars
make the Army more or less competent in its handling of small wars? In
doing so, this chapter compares the operations of the Indian and British
armies in Burma during the Second World War with those undertaken by
the British Army during the Malayan Emergency. The comparison is apt not
only because the Malayan Emergency followed within three years of the end
of the Burma campaign, but also because the jungle terrain in both theatres
required the use of similarly unconventional tactics.

The experience of the British Army in the Far East early in the Second
World War was similar to that of the first years of the Malayan Emergency.
Neither the Indian/British Army of 1941 nor the British Army of 1948 was
prepared to fight in the way that the campaign required. This was because,
first, the troops involved had received minimal conventional training in
1941, because of overexpansion, and in 1948 because of the rundown in
veteran troops and their replacement with National Servicemen. Second,
even had troops been properly trained for conventional (or open style)
warfare, no centralised systems had been established to train and promote
tactical doctrine for jungle warfare. Jungle warfare is a specialised form that
is essentially an adaptation of conventional tactics, but one that requires
dedicated training. As the 1943 Indian Army training pamphlet, *The Jungle
Book*, stated: 'there is nothing new in jungle warfare [tactically speaking],
but the environment of the jungle is new to many of our troops. Special
training is therefore necessary to accustom them to jungle conditions and to
teach them jungle methods'.[1]

The Indian and British forces that fought in the Malaya and Burma
campaigns of 1942 were unprepared for the realities of warfare in the jungle.
The defeats of 1942 and 1943 motivated the Indian Army to reform itself
tactically and implement an organisational system to respond to the need for
appropriate training. By 1944, all units in India Command that had been
earmarked for service against the Japanese received training at jungle
warfare schools or training divisions, following common doctrine that had
been established in September 1943. Indian and British units implemented

an intensive programme for the ongoing assessment and reform of tactics as battlefield demands shifted. By the end of the war, the Indian and British armies boasted a sizeable force of jungle trained, battle conditioned veterans. Many of these men, however, returned at the war's end to homes in India, Nepal and, to a much lesser extent, the United Kingdom.

One further development from the Malaya and Burma campaigns of the Second World War that had an important connection to operations during the Malayan Emergency was the importance of junior leadership. *The Jungle Book* comments on this also:

> Experience shows that command must be decentralised so that junior leaders will be confronted with situations in which they must make decisions and act without delay on their own responsibility. The ability to make sound decisions can only follow from thorough training and continuous practice.[2]

Due to the terrain in both campaigns, junior officers and senior NCOs found themselves responsible for making important command decisions that their counterparts in Northwest Europe or in post-war Germany had not been faced with. India Command's Infantry Committee Report of 1943 and the findings of Malaya Command both stressed the need to properly train and prepare junior officers and NCOs for the responsibilities of decision making while on patrol.

The end of the Second World War and the distraction of other commitments for the British Army meant that many of the lessons and doctrines that had been developed in the Second World War were forgotten for a short period of time. Most units, even formations stationed in the Far East, carried out training exercises which focused on open style warfare and tactics of aid to the civil power. When the Malayan Emergency was declared in 1948, senior commanders were largely unaware of the need to train their men in jungle warfare tactics and to create a simple doctrine for them to follow. As had happened in the Second World War, initial efforts to address the issues particular to jungle warfare were scattered and uncoordinated. Unlike the Second World War, however, Malaya Command did have resources at its disposal – a vast amount of written information, plus the firsthand experience of veteran officers.

Burma and Malaya: the Second World War

The British and Indian Army was beaten in both Malaya and Burma in 1942 by superior tactics and a better trained Imperial Japanese Army. The British, Indian and Australian units were road bound, and poorly equipped to counter Japanese envelopment tactics. The Imperial Japanese Army was more mobile and had been trained to manoeuvre and fight in the jungles of Southeast Asia. Even with Japanese jungle warfare training centres established in Formosa and Hainan, the Japanese units that fought in the Malaya and First Burma campaigns were not experts in jungle warfare, but their basic understanding of what would work far surpassed that of their opponents.[3] The essential feature of Japanese tactics was that the attacking formation was divided into three groups. The first group would attack the enemy position from the front, while the second group would move around the flank of the enemy, either to attack from the side or the rear, or to form a roadblock to the rear of the enemy's position, creating potential panic amongst untrained troops who feared being cut off. The third group was left in reserve to support the frontal attack, flanking attack or rear position.[4]

The British and Indian Army units that were stationed in Malaya and Burma were prepared neither for this type of warfare nor for a fight with the Japanese. The Indian Army was rapidly expanded from 1940 onwards; many units were milked of experienced officers, Viceroy Commissioned Officers and men, who were reassigned to newly raised units. These units were trained for 'open style' warfare, and expected to be sent to North Africa. Many of the British Army units that were serving in India, Burma or Malaya were also below full strength, as experienced men sought transfer to units being raised or fighting in Europe.[5] These units, too, for the most part trained in open style tactics.[6]

It was unquestionably important to train the recruits in the basics of conventional warfare. The men and officers needed to have a firm footing in these basics before taking on specialised training such as jungle warfare. However, the rapid expansion of the Indian Army and the milking of regular British Army units prevented many units from reaching the standards required for even conventional open style warfare. Training was hampered further by delays in the deployment of new weapons; the forces that were to serve in Malaya and Burma were not considered a high priority. This makes sense considering the requirements of outfitting troops in the UK, North Africa and the Middle East. Nevertheless, the fact remains that many units were not given Bren guns or 2-inch or 3-inch mortars in sufficient time for the troops to train with them.[7]

British and Indian units had served in 'bush' operations since the end of the nineteenth century, but there was no major dissemination of information based on past experience that might be relevant to fighting in the jungles of Burma or Malaya.[8] The first major work created to address this area was the Military Training Pamphlet No. 9 (India) in 1940, the first of several editions and later known as *The Jungle Book* – although it is interesting to note that the header for the first edition specifies 'Forest and Jungle Warfare'.

There was one major attempt, before the Japanese invasion, to train troops stationed in Malaya in jungle combat techniques. Lieutenant Colonel Ian Stewart, commander of the 2nd Battalion, Argyll and Sutherland Highlanders, decided to devise tactics and train his battalion. Over the course of two years, he and his officers worked through various exercises, anticipating the potential problems of operating in the jungle. He set out to make the jungle a liveable environment for his men, and identified the need for active patrols through the jungle, rather than along the road. He devised new methods for all-round defensive positions, from which the enemy was kept at bay by aggressive defensive patrols.[9]

For the most part, Stewart's initiatives went unnoticed. His battalion was part of the 12th Indian Brigade, and there is evidence in unit records that other units within the brigade followed similar training. These records also indicate that the constant 'milking' of troops created problems; training had to be done over and over again as new men arrived and veteran troops withdrew.[10] Most of the other units stationed in Malaya did not undertake any jungle warfare training.[11] When the 12th Indian Brigade was forced to withdraw, with other elements of the allied forces, following the successful Japanese invasion, elements of the 2nd Argylls, including Lieutenant Colonel Stewart, were able to get to India. They brought with them experience and ideas about fighting the Japanese in the jungles of Malaya.

British and Indian Army units also attempted to develop strategies for fighting in the jungle during the first Burma campaign, but they had very little time to assess their situation. Some units recognised that a specialised training scheme was needed but did not have the time to develop one. As units withdrew towards India in the wake of the Japanese invasion, they did attempt to get off the roads or go around Japanese roadblocks, but ultimately they were not given time to regroup and work out possible alternative tactical procedures.[12] Unlike many of the forces stationed in Malaya, units of the 1st Burma and 17th Indian Divisions were able to withdraw into India, bringing with them a large cadre of men and officers who had experienced the tactics of the Japanese and of operating in the jungles and open plains of Burma.

Jungle warfare assessment in the Far East, 1942–5

Even before the Malayan campaign had ended, some of the veterans who had escaped were discussing lessons to be learned from its outcome.[13] The end of the first Burma campaign in May 1942 added more voices and more information to the process. The thirteen months between its conclusion and June 1943 were a period of intensive discussion and tactical reform. As had happened before, these reforms were initiated at the unit level, so there was no consistent programme. The defeat the army suffered in the first Arakan offensive in June 1943 finally convinced GHQ India that appropriate and consistent training should be a priority. This was undertaken with the establishment of training divisions and pamphlets on tactics of jungle warfare, and implemented with the help of soldiers trained in jungle warfare.

Reform: May 1942–June 1943

Some senior officers believed that a major reason for the defeats in the Malayan and first Burma campaigns was a lack of offensive spirit among the allied forces. Senior officers such as Lieutenant General William Slim, Major General D.T. Cowan and Lieutenant General Reginald Savory discounted this theory and stressed instead a need to devise new tactics to engage the Japanese. Officers who had combat experience against the Japanese were interviewed, and their reports passed on to GHQ India for dissemination in Military Training Pamphlets (MTPs) or Army in India Training Memoranda (AITM). Officers and men were also sent to lecture throughout India Command on their experiences.

During the initial period of reform, from May 1942 to June 1943, the process of implementing reforms was very decentralised. Specific manuals dealt with jungle tactics, but the vast majority of training material available up to mid-1943 focused on open style warfare. Additionally, Indian and British Army units in India Command continued to suffer a drain of manpower, with many veteran units below strength as a result of campaign losses and disease problems. The political disruption of the 'Quit India Movement' of August 1942 created further disruptions in training when units were called out to aid the civil power. As a result there was no way to assess whether information provided to units in the field was actually being used.

AITM No. 15, published in March/April 1942, laid out many of the fundamentals of jungle warfare training, beginning with the assertion that 'tactics of jungle warfare are specialised and to employ them well special training is needed'.[14] Other essential principles outlined include: the importance of proper patrolling to gain intelligence; the impracticability of linear defence and the superiority of all round defence; and the need for all the arms of the army to be trained to fight as infantry in preparation for jungle conditions.[15] Subsequent AITM incorporated lessons learned from subsequent campaigns.[16]

Some of the first steps towards a centralised training structure began at the divisional level. For example, the 14th Indian Division created a jungle warfare school at Comilla in mid-1942,[17] although it did not have the time or resources to retrain all the units under its command. It was still in the process of incorporating units that had just been formed and lacked even 'open style' warfare training. Three other divisions undertook similar initiatives;[18] the initiatives undertaken by the 17th Indian Division were characteristic.

The 17th Indian Division[19] had fought throughout the first Burma campaign, and its commander, Major General Cowan, was quick to recognise that his division had been outfought and to try to find out how. One of his brigadiers, R.T. Cameron, 48th Indian Brigade, wrote a report detailing the failings of the first Burma campaign and outlining possible solutions. The division used this information to create its first Training Instruction in June 1942. The document described how many officers and men felt:

the division has acquired considerable practical experience of fighting against the Jap [*sic*] and many lessons have been learned from their methods which can be adopted by us. ... now is the time to train and practice these new methods and to drive in the good lessons before they are forgotten.[20]

Cowan made practical decisions early on. He also recognised that many men and officers were intimidated by the jungle environment and ordered small base camps to be set up in the jungles surrounding the Imphal plain. From these the men could operate in small patrols, and acclimatise themselves.[21] In the second Training Instruction, Cowan directed that all men and officers, before being allowed to go on leave, would be required to record their experiences and observations and submit them to their respective headquarters. This information formed the basis of lectures and notes that were to be distributed to newly arrived replacements.[22]

Following on from initiatives undertaken at the divisional level, GHQ India decided to re-organise divisional structures[23] to facilitate operations in the jungles of Burma. They created a new type of mixed transport division named the Animal and Motor Transport (A and MT) division[24] in October 1942.[25] The rationale for this was that all units were to be a mix of mule and lorry transport. Animal transport would be in the first echelon, moving with troops, and the mechanised element would be the second echelon. This new establishment was intended to get units off the roads and free them from their baggage.[26] Over the course of the following two years, this divisional organisation (low mechanised) formed the basis for most of the divisions serving against the Japanese.

In January 1943, GHQ India ordered the 7th Indian Division[27] to re-organise along the lines designed by the new A and MT specifications, and to serve as the first purpose trained jungle division. The division went through all the required organisational reforms, and reported its findings to GHQ India for further revision.[28] All units also undertook intensive jungle warfare training, and attended lectures given by veterans of the Malaya and Burma campaigns.[29] The battalions carried out battalion-sized, and later brigade-sized exercises. Divisional conferences were held to discuss problems and lessons learned over the training period.[30] In May, the division was ordered to move to Ranchi,[31] in preparation for deployment to the Arakan region.

While the 7th Indian Division underwent this training process, two events occurred that changed the drive for proper jungle warfare training throughout India Command. These were the first Arakan offensive and the first Chindit operation.

A limited offensive was ordered in the Arakan region for December 1942. The very fact that the operation was launched highlighted the lack of a cohesive strategic and tactical plan for fighting the Japanese in the jungles of Burma. The 14th Indian Division was earmarked to attack; although it had created a jungle warfare school, many of the men and officers had not been trained properly when the division was committed to battle. This issue was compounded by the fact that any reinforcements lacked not only jungle warfare training but even basic conventional training.[32] To make matters worse the division was still operating on an establishment of transport that relied heavily on road communications; the new A and MT division establishment had not yet been implemented for the army as a whole.

Nine brigades[33] in all were involved in the attempt to break through at selected points in the Donbaik and Buthidaung regions. As early as 14 January, the Commander-in-Chief of the Indian Army, Field Marshal Sir Archibald Wavell, stated 'we still have a great deal to learn about jungle fighting'.[34] By March, the Japanese launched a counter-offensive. The Japanese were successful in cutting off the road communications north of the 14th Division's positions by early April. After fierce fighting, units of the 14th Division had withdrawn to their original lines by early May, a much demoralised force. A major centralised training and reinforcement procedure was needed, and the actions of the 14th Indian Division made this clear to all in GHQ India.

The first Chindit operation has been the subject of much examination. It may still be open to debate whether the operation was successful in operational terms, but it is beyond question that it accomplished two important feats. First, it raised morale among the British and Indian forces, which was sorely needed after the debacle of the first Arakan. Second, it served as a testing ground for tactics that would become the hallmark of operations in Burma, particularly re-supply by air.[35]

Centralised reform: June 1943 – operations Ha-Go and U-Go

As a result of the debacle in the first Arakan campaign, India Command decided to centralise and focus on jungle warfare training and reinforcement structures. Field Marshal Wavell, in one of his final acts as Commander-in-Chief, India, ordered the formation of a representative committee to examine recent combat performance and make recommendations for improving the infantry and strengthening the overall morale of the army.

The Infantry Committee was formed in response to Wavell's instructions, and its report considered a broad range of issues affecting the troops of India Command. Some of the problems that the report identified included a lack of basic training for most units and a considerable level of inexperience among the officer ranks. Potential deployment of troops to either the Far East or the Middle East added further complications to training and establishment structures. The committee's main recommendation was for a simple, consistent and officially recognised jungle warfare doctrine. GHQ India was instructed to take responsibility for producing and disseminating written documentation in support of this doctrine.[36]

The report also provided suggested timetables for basic and jungle warfare training, and recommended the establishment of two training divisions and, ultimately, a training brigade.[37] Under the recommended structures, basic training was extended from three to a minimum of six months. Basic conventional training would take place at regimental centres and focus on weapons training, discipline, individual, section and platoon training.[38] Structures would be established to track training for regiments being sent to other theatres separately from those destined for jungle combat. Following basic training, officers and men would be sent for two to three months of jungle warfare training at one of the two designated training divisions or brigade.[39]

Recommendations made in the report were implemented almost immediately; by the end of 1943, both training divisions had been established. Both divisions attempted to have as many veteran officers and VCOs as possible on hand to create the training regime and oversee the establishment of the training grounds. The informational pamphlets published by GHQ India formed the basis of the training programme, supplemented by the personal experience of the trainers.[40] The training programme was consistently updated over the course of the war as lessons were learned from the fighting in 1944 and new documentation was prepared.[41]

General Claude Auchinleck took over as Commander-in-Chief in late June 1943. He was an enthusiastic proponent of the report's recommendations from the beginning and worked to ensure the smooth progress of the training programme. The establishment of South East Asia Command also meant that India Command became the main training and supply base for the war in Southeast Asia, while strategy was decided elsewhere. As part of this reorganisation, the 14th Army was established and placed under the command of Lieutenant General William Slim. Slim, like Auchinleck, recognised the need for the army to retrain. He had himself written reports after the first Burma campaign, suggesting new tactics to use against the Japanese.

As a result of the reforms implemented in 1943, the troops earmarked for duty in Assam and the Arakan were trained to an unprecedented level on the eve of the Japanese offensives of 1944, Ha-Go and U-Go. Three kinds of division had been earmarked for duty in Burma: A and MT (high scale),[42] A and MT (low scale)[43] and a light division.[44] Reinforcements were being trained at the two training divisions and the brigade. A common doctrine had been developed, and disseminated through *The Jungle Book*.[45] Divisional commanders used the various pamphlets available as a base from which to draw lessons and information. Most divisional commanders, like Cowan of the 17th, created their own specialised training instruction and circulated it in pamphlet form. So little information was available initially that pamphlets such as these became a primary source of tactical training instruction. The process of assessment and development of new tactics was ongoing; as units came into contact with Japanese patrols, they reported on their experiences so that decisions could be made about changes to the training structure.[46]

Jungle tactics used in Burma

The chief element of tactical success in the jungle was continuous patrolling. Due to terrain, there were large tracts of no-man's land between the Indian and British forces and the enemy. The deployment of 'active' patrols was, as *The Jungle Book* stated, 'to make no man's land your man's land'.[47] Intelligence gathering was a considerable problem during the Burma campaign, mainly due to the difficulties of operating in the jungle terrain. The usual methods of gathering information were almost completely unavailable. Reconnaissance (recce) patrols were the most effective solution that the British and Indian Army designed to identify Japanese units and their intentions, as these could move through the jungle without alerting the enemy to their presence.

The box[48] was the chief defensive formation used by units, first in the jungles and later in the open plains. Each formation, regardless of size, established its position in this way, digging slit trenches and securing its perimeter with fire lanes and patrols. The box was intended to encompass an active defence in all directions, and to be constructed and defended well enough to operate on its own or in support of another box under attack. For the most part the box was structured as follows: half the troops were responsible for perimeter defence. One quarter of the force was held in reserve, to be deployed to counteract any Japanese breach of the wire. The final quarter was used outside the perimeter to destroy the Japanese as they formed up for an attack. The troops were required to maintain fire discipline and keep movement to a minimum in the perimeter. The boxes were to remain put if the lines of communications were cut, and could expect to be re-supplied by air,[49] while a larger relief force was sent in to destroy the Japanese positioned outside the box perimeter.

Patrols would be sent out from the box to secure the area around the position. When patrol bases were set up, two kinds of patrols were sent. The first were small reconnaissance patrols who reported on Japanese movements or positions. The reconnaissance patrol consisted of three to four men. If the reconnaissance patrols reported Japanese movements or positions, the company commander or battalion commander would decide to launch the second kind of patrol, the fighting or tiger patrols. The size of fighting patrols ranged from no less than a platoon to as many as a company. They laid ambushes or attempted to destroy Japanese patrols or positions. All the men were to be trained in jungle lore and minor tactics. They were not to follow jungle tracks if they could avoid it. Bases were not to be set up on a jungle track but in places that overlooked a jungle track. Rendezvous were set up for every patrol; in case the patrol was ambushed, all the men would be aware where to go. The main goal was always to outflank the enemy and not to allow oneself to be outflanked.[50]

In terms of offensive action, attacks were to be divided into three phases. In the first phase reconnaissance patrols were sent out to assess the enemy's flanks, his position and depth of defence. The second phase was a night advance towards the enemy's position, which would attempt to get behind the rear of the flank. All patrols would dig defensive boxes in case the Japanese counter-attacked. Fighting patrols would be launched to clear Japanese patrols that might be lurking. The third and final phase consisted of large flanking attacks. Some men were left opposite the Japanese positions to hold down their front, while fighting patrols attempted to cut off the lines of communication. All these strategies were designed to avoid launching frontal attacks, but this was not always possible due to terrain or time constraints.[51]

The battalion used successive attacks to move forward. As each position was taken, the men would immediately consolidate to prepare for a potential Japanese counter-attack. As forward areas were cleared of enemy positions, the track or road behind the battalion positions was reinforced to facilitate delivery of supplies to the forward positions. The battalion also used boxes to protect the tracks or roads, although, as was seen in 1944 and 1945, if the roads or tracks were cut, air re-supply was used instead.[52]

The Japanese were decisively defeated during the fighting in 1944. Air re-supply was a significant factor in this turn of events, as has often been contended, but more important was that the units involved fought in a different manner than the Japanese had expected. The troops were properly trained to fight in the hilly and jungle-covered terrain of the Imphal area and Arakan peninsula. The engagements of 1944 were an ongoing learning process and some units encountered difficulties initially, but all the units deployed in Burma were able to hold their ground when ordered and carry out successful counter-attacks that eventually forced the Imperial Japanese Army back towards Burma. Japanese officers who were interviewed after the war commented on the change. Lieutenant Colonel Fujiwara, a staff officer with the 15th Army, stated that the Japanese reversals of 1944 were due to 'failure of recognition of the allied equipment and training in jungle warfare'.[53] When discussing British/Indian 'box formations', a Japanese colonel commented that 'until this time [1944] our forces did not encounter this type of enemy tactics. ... and therefore were unprepared to deal with them'.[54]

Post-battlefield assessment and training: autumn 1944–February 1945

Even with the successes of 1944, British and Indian troops recognised that there was still room for improvement as fighting in Assam and the Arakan began to wind down.[55] GHQ India published assessments of engagements in AITM and MTPs that were distributed to formations of 14th Army. As divisions were pulled out of the line to reinforce, they were instructed to retrain their men in preparation for the 1945 campaign.

The 220 Military Mission Report was the source of much of the continuing improvement in tactics and training. This mission, which was headed by Major General J.S. Lethbridge, British Army, had been instructed to report to jungle warfare centres and formations throughout India Command, Australia, the USA, the Solomon Islands, New Caledonia and New Guinea.[56] His report was intended to serve as the basis for a training programme for British units sent to the Pacific theatre as the war in Europe came to an end.[57] The War Office in London may have been hesitant about initiating change, but Lethbridge was not. When he submitted his report in March 1944, he specifically stated that British units re-assigned to the Pacific theatre needed to become jungle orientated and adopt the various A and MT organisational structures already in use by India Command.[58]

A meeting was held in May 1944, with representatives from India Command, 11th Army Group, and officers of the IV, XV, XXXIII Corps, to discuss the report and how to implement its findings. Lethbridge's[59] recommendations included the creation of a standard divisional structure[60] for South East Asia Command (SEAC). The proposed division was expected to be capable of jungle fighting, undertaking air transport, and making amphibious landings. All divisions were to be assigned a divisional headquarters protection battalion,[61] and a reconnaissance battalion was to be attached to each division, strictly as 'light infantry', not a mixed force of mounted and mechanised. A medium machine gun battalion was to be attached to each division as well,[62] and artillery was streamlined.[63] Organisational changes would be implemented as each division was pulled out of the line for reinforcements and retraining.

The dissemination of information gathered in fighting continued to be written up and distributed through the AITM, MTPs and Director of Infantry, Monthly Pamphlets. These documents reinforced a consistent message on a few key points: patrolling, defensive boxes, track discipline and camouflage. AITM nos. 24 and 25 highlighted lessons from engagements in Assam and the Arakan, as well as incorporating perspectives from American and Australian forces in the Southwest Pacific.[64] In early 1945, GHQ India published a new, comprehensive jungle manual, the *Jungle Omnibus*. It was intended to supplement, rather than to replace, *The Jungle Book*. It describes its own mission thus:

> This omnibus contains everything that has been published on jungle warfare in AITM during the last two years. It is issued for the guidance of new units and formations who may arrive in India from overseas without previous knowledge of jungle warfare training, and with very few pamphlets on the subject.

Distribution was intended to be one copy per officer for units arriving in India from overseas,[65] but the *Jungle Omnibus* also made it into the hands of numerous veteran officers.[66] When the War Office in London focused its attention on developing documentation for jungle warfare training, it created Military Training Pamphlets nos. 51 and 52; the latter is essentially a copy of *The Jungle Book*.

All of the divisions involved in combat in 1944 undertook retraining in late 1944 and early 1945.[67] Formations which had not seen active service, such as the 19th Indian Division, were re-organised and carried out further training based upon the lessons of the 1944 campaigns.[68] Examination of the war diaries of the 5th, 7th, 17th, 19th, 20th, and 26th Indian Divisions, as well as individual units, provides extensive evidence of training instructions and courses.[69]

One battalion's experience of retraining is worth particular consideration. The 4/8th Gurkha Rifles, 89th Indian Brigade, 7th Indian Division, had taken part in jungle warfare courses at Chindwara in 1943, and then seen service in both the second Arakan offensive and the fighting on the Imphal plain. Lieutenant Colonel Walter Walker, previously second in command, took over command of the battalion after Imphal. Walker is notable because he played a central role in the development of a jungle training doctrine that continues to influence British Army practices to this day.

Units of the 7th Indian Division were assigned to carry out re-training exercises in jungle warfare during October and November 1944. Previous to this the divisional commander, Major General Frank Messervy,[70] had developed a comprehensive plan for re-training the division,[71] and this had been supplemented by a series of instructions prepared by 89th Brigade staff and entitled 'Lessons from Operations'.[72] Walker, meanwhile, had been busy creating his own 'lessons from operations', drawing heavily on AITM nos. 24 and 25.[73] Walker encouraged his officers to discuss recent engagements and lessons to be learned from their experiences. He believed that everyone could learn something, even veteran officers.[74]

Training for the 7th Indian Division began with jungle craft and proceeded through platoon, company, battalion, brigade and divisional exercises. Reinforcements arrived from the 14th and 39th Indian (Training) Divisions.[75] Veteran officers commented that the standard of training was high, but needed refining. The course of exercises helped units to mesh completely. Officers felt that this programme brought the 4/8th Gurkhas to a level of training considerably higher than when they had first left Chindwara in the summer of 1943, and most credited Walker's drive to retrain all the officers and men for this development.[76]

To summarise, the learning process within India Command was slow initially; it took time for senior officers to recognise the need for specialised training. Rapid expansion of the Indian Army and political problems within India created further complications in the early years of the war. Thanks to the efforts of a number of senior and junior officers, awareness of the need for reforms in doctrine and training began to grow, and the whole of India Command was convinced after the disastrous first Arakan operation of 1942–3. From 1943 to the end of the war, units sent to Burma were trained to the level and specifications required for the campaign. Both Indian and British units continued to assess lessons and apply the results to their efforts in training both veterans and raw recruits through 1944 and 1945. By the end of the Burma campaign, the units[77] stationed there had developed a high level of expertise in opposing the Japanese forces. They had successfully adapted to the changing conditions of Burma, both on the open plains and in the jungle.

Perhaps equally important to the victory achieved over the Japanese was the accomplishment of mastering the techniques necessary to succeed in jungle operations. Overcoming their initial fear of manoeuvring in the jungle, the British and Indian units were eventually to become seasoned experts. As a result of this success, the British Army attained a level of confidence in their capabilities that would serve them well when the time came to confront the problems in Malaya.

The Malayan Emergency

The Malayan Emergency was a smaller conflict than the Burma campaign, with much lower numbers of troops deployed. The enemy, the Malayan Races Liberation Army (MRLA), also known as the Communist Terrorists (CTs), was a different kind of enemy from the Japanese. They were jungle fighters, but had more in common with guerrillas, as they tended to ambush the security forces (SF) when they were in superior in number to the SF. They were excellent ambush soldiers, but tended not to have the same tenacity in defending their positions as the Japanese had demonstrated. Brigadier Dennis Talbot, lecturing to New Zealand officers in 1950, put it succinctly:

> there is one vital difference in present operations to what we learnt in jungle warfare against the Japanese ... Japanese organised troops who held position and lines of communications ... were ready to and willing to fight. ... [B]andits are guerrillas who depend on hit and run tactics. ... [T]heir object is to avoid clashing with the security forces except when the latter are greatly outnumbered and taken by surprise. ... [T]hey do not defend or hold any particular area or function on a line of communication.[78]

The victory in Malaya rested on a combined political and military strategy to counter the insurgency.[79] The SF were one piece of a plan to defeat the insurgents in society as well as in battle.[80] As the *Conduct of Anti-Terrorist Operations in Malaya* (*ATOM*) training manual stressed in the opening paragraph of 'Own Forces': 'the responsibility for conducting the campaign in Malaya rests with the Civil Government. ... [T]he Armed Forces have been called in to support the Civil Power in its task of seeking out and destroying armed Communist Terrorism'.[81] The *ATOM* further stressed that

> the primary role of the Army is to seek out and destroy CT in the jungle and on its fringes. By the constant harassing of the CT, their lines of communication with sympathisers amongst the civil population are

disrupted. Thus in an effort to maintain their food supply system they are forced into the open and so brought to battle.[82]

Initially, senior military commanders viewed the campaign against the MRLA as a conventional conflict, centred on seeking out and destroying enemy formations as they had in Europe. Many did not realise at first that troops would need specialised training in order to operate effectively in the jungle. In 1976, Lieutenant General Walter Walker recalled that, at that time,

> the fundamental trouble was that within two years of defeating the Japanese in Burma, all our military training and thinking had become focused on nuclear and conventional tactics for a European theatre. … so when the Malayan Emergency broke out, we had forgotten most of the jungle warfare techniques and expertise, learned the hard way at such cost in the Burma Campaign.[83]

It would be necessary to revisit the lessons of the Burma campaign, because the reality of the war against the MRLA, as it had been partially against the Japanese in Burma, was an endless series of patrols, punctuated by occasional contacts or ambushes. Just as their predecessors had in Burma, the troops that fought in Malaya had to learn to master the techniques of the jungle in order to deny their enemy supremacy in the jungle. This realisation was slow in coming, but once recognised became the cornerstone of tactical developments in Malaya. Units in Malaya were thoroughly trained for the terrain in which they were operating, with emphasis on ongoing assessment and implementation throughout Malaya Command and the Far East Land Forces (FARELF). The officers making decisions about doctrine had a wealth of material and personnel to draw upon in developing jungle warfare tactics for the troops, thanks to the earlier experiences and documents of the Burma campaign.

The period from 1945 to the start of the Emergency in 1948 was a difficult time for proponents of jungle warfare. British involvement in the Indian Army ceased in 1947 with the declaration of independence for India and Pakistan. Many Burma veterans from British units had been released from service, and had returned to civil life. Units that were initially deployed to FARELF were below strength and poorly trained, even for a conventional conflict.[84] Several Gurkha regiments were included in the FARELF deployment; they too were in a state of flux, in the midst of a major re-organisation following their release from the Indian Army.[85] The inception of the MRLA campaign caught British and Gurkha units off guard.[86]

Re-learning the tools of the trade: 1948–50

The first attempts to address the particular issues involved in jungle warfare began within a month of the Emergency being declared. These early initiatives took shape in the creation of the 'Ferret Force' in July 1948.[87] These units were to operate as six small, 15-man units, made up of Gurkhas, British soldiers, Malayan Police and liaison officers. Many of the experienced officers came from Gurkha battalions and Force 136, a covert unit that had served alongside the MRLA during the Japanese occupation of Malaya in the Second World War. Lieutenant Colonel Walter Walker, of the 1/6th Gurkha Rifles, was to train the new units.[88]

After a month's training the first units were ready. The Commander-in-Chief, FARELF, General Sir Neil Ritchie, and Lieutenant General Sir Charles Boucher, GOC Malaya Command, reviewed an exercise with Ferret Force, and were suitably impressed. Following this review, the first units were deployed for long jungle patrols, where they operated successfully. Despite this, the units were disbanded at the end of the year, as a result of control issues that had developed from mixed units of police and army troops. Even so, the groundwork for specialised training had been laid.[89]

In the wake of the Ferret Forces' success, General Ritchie decided that all units deployed from the United Kingdom should receive appropriate jungle warfare training.[90] In order to implement this initiative, he decided to appoint Walker as the new head of the FARELF Training Centre (FTC), or Jungle Warfare School.[91] Walker immediately set about creating a jungle training programme similar to those that had been developed in India a few years before, even using the tactical pamphlets produced for the Burma campaign. In 1949, Walker commented that

> the amount of real abysmal ignorance on the subject of jungle warfare that exists amongst junior officers and NCOs is beyond belief. ... [O]ne can only assume that they have not studied or been made to study the following most useful pamphlets, MTP No. 51 and 52 and The Jungle Book 1943.[92]

He gathered officers and NCOs with jungle warfare experience, and they created a training pamphlet of their own. Prospective instructors were put through the course before the first troops arrived.[93]

The overall plan for how training would be carried out was different from the programme developed by India Command. Instead of creating a training division or brigade, the school was to be the central mechanism for carrying out training. When battalions were called up in the United Kingdom for service in Malaya, a small cadre of officers and NCOs were sent to the FTC. Battalions already stationed in FARELF[94] also sent cadres of officers and NCOs to the school.[95] The course was structured over

> 167 hours of instruction, 110 given over to 4.5 days in the jungle, 24 to a preliminary day in the jungle, 16 to immediate action drill, 8 to jungle navigation, 4 to MT ambush, 3 to observation and tracking, and 2 to jungle marksmanship.[96]

Officers or NCOs who paid attention were expected to complete the course within three weeks.[97] The school ran nine courses a year, each consisting of thirty-six officers and NCOs. Trained cadres would return to their battalions, which were stationed near the FTC or another convenient location, and run a jungle warfare course for the rest of the battalion. This requirement was carried out with varying degrees of success.[98] Instructors at FTC worked to keep pace with developments and amendments in the field, and to update their training regime accordingly, following a practice that had been established in India Command.

The first quarterly report of the FTC makes clear that Walker and his staff understood the problems that they faced in preparing units for jungle combat. As was the case with troops preparing for Burma in 1942 and 1943, many officers and NCOs deployed to Malaya were lacking in basic tactical skills and training. New arrivals apparently had little idea of the basics of tactical organisation for moving through either jungle or open terrain. Many trainees did not understand that the basic battle craft organisation (three groups: a reconnaissance group, a supporting group (Brens), and an assault group (riflemen)) was suitable for all operations in open, as well as jungle, terrain. Walker and his staff commented that: 'many officers and NCOs [have] minimal knowledge of minor tactics. ... many feel they know all that is needed to fight in the jungle'; the term 'jungle bashing' had crept into the lingo; 'men need to be a poacher not an elephant in the jungle'; and finally when the standard of training was compared to that of units in Burma in 1944 and 1945, 'it is absolutely clear that we have a long way to go'.[99]

Walker and his staff emphasised that small reconnaissance patrols, similar to those used in Burma, should be sent out to gain information before larger fighting patrols were sent out to engage CTs. Walker stressed that the ability to send out good reconnaissance patrols required a number of skills: excellent deceptive abilities; discipline; a sense of direction in the jungle, day and night; and concealment.[100] While trainees were learning this doctrine, battalions already in the Malayan jungle were operating in large-scale sweeps through the terrain. They were often unsuccessful in finding and destroying MRLA troops, who could hear them coming, lay an ambush, and withdrew.[101]

Continuing to draw on his experience in India, specifically with his old regiment, the 4/8th Gurkhas, Walker encouraged discussions regarding tactics and other issues. Some officers contended that the responsibilities they would have when they returned to the battalion would prevent passing on all the knowledge that they had gained to others in their units.[102] This last point was, in fact, a significant problem, and was likely to be directly affected by the attitude of the CO towards jungle warfare training.[103] As a result, it took some months to resolve effectively. Walker, commenting on this issue, specifically stated in the report for April to September 1949 that

> although the school has trained 711 officers and NCOs in jungle warfare during the past 12 months this effort has concentrated little towards the common object because officers and NCOs have not been given the opportunity to spread the gospel. … instructors on return from tour invariably comment on the appalling low standard of elementary tactics that exist.[104]

CT activity increased measurably in the later months of 1949. This created a gap in the jungle warfare training programme, as all officers and NCOs were needed within their battalions. During this interval the instructors went into the field to assess the abilities of the troops, while Walker returned to the 1/6th Gurkha Rifles and the FTC was taken over by Lieutenant Colonel J.H. Law.[105] The FTC continued to serve as the main training ground and tactical doctrine development centre for troops arriving into FARELF for duty in Malaya. The CTs remained elusive throughout this period, inflicting heavy casualties on the civilian population and maintaining control of the jungle. Many British and Gurkha units had still not gained the knowledge of jungle warfare needed to gain operational superiority over the MRLA.[106]

1950–2: formal doctrine formulated

Strategically speaking, the conflict in Malaya took a fundamental turn for the better in 1950 with the creation of a new post, that of Director of Operations. This post was intended to streamline the efforts of the Malayan Police and the Army, eliminating duplication of effort. The man appointed to fill the position initially was retired Lieutenant General Sir Harold Briggs, a veteran of the Burma campaign.[107] Under Briggs' leadership, a cohesive strategic plan took shape. Known as the Briggs Plan, its chief element was re-settlement of villages in protected areas. This idea did not originate with Briggs, but he took the initiative to implement it, in concert with a number of others. The Briggs Plan had several key elements: village re-settlement; denial of food to guerrillas; and cooperative intelligence between the police and the army. The principal role for the British Army under this plan was to get into the jungle and act as a barrier between guerrillas and the civilian population, cutting off the guerrillas' access to food, support, and intelligence from the local residents.[108]

The development of cooperative intelligence procedures between the police and the Army was a necessary part of the plan because it enabled the command leadership to gather information about the civilian population as well as the guerrillas. A command and control structure was established at national and local levels that incorporated members of the police and military hierarchies in order to coordinate and corroborate information gathered from a variety of sources. This structure was an important development in counter-insurgency doctrine,[109] and would become a standard for future operational practice.[110]

In addition to implementing this strategy, the British Army had decisions to make about how best to deal with the MRLA militarily. By 1950 the MRLA still had not been decisively defeated. They had become very elusive, but continued to carry out ambushes and attacks effectively. The large-scale operations that the army had used to try to destroy the MRLA had not achieved their objectives.[111] As had happened during the Burma campaign, individual battalion commanders independently developed and tried different plans, most of which involved carrying out small reconnaissance patrols, supporting fighting patrols, and similar methods which were more effective at disrupting the CTs and killing quite a few.[112]

As Briggs arrived in Malaya, General Boucher was relinquishing command as GOC Malaya because of illness. Major General Roy Urquhart took over command. He was initially supposed to be a temporary replacement,[113] but he immediately got involved in the various debates ongoing in the Army on issues of tactics and strategy, and in July 1950 called a major conference to address them comprehensively.[114] The stated objective of the conference was 'to study and discuss the army's problem in Malaya and to devise the best means of dealing with it. ... [K]nowledge gained by experience of actual operations can be pooled from for benefit of all.'[115]

Topics discussed at the conference ranged from how long units should remain in the jungle to what types of training should be offered at the FTC. The FTC was to continue in its capacity as the army's main training ground, but was to be moved to Kota Tinggi. Lieutenant Colonel Michael Calvert gave a lecture on issues relevant to patrols, strongly recommending that patrols be properly debriefed of their information. He also stressed that units needed to get behind the bandit screen of information and that to do so, they must be able to slip into the jungle without being detected. Finally, he called for the formal establishment of a unit trained for deep jungle operations and re-supplied by air. This proposal was approved,[116] and the Malayan Scouts were born.[117]

Two major debates on matters of strategy were presented during the conference. One focused on the optimum size of operations and patrols in the jungle. The final decision mandated small reconnaissance patrols with appropriate backup from strong fighting patrols. This ruling meant a shift in how a company would operate as the main tactical formation in the jungle. As a result, the FTC was asked to create a separate course specifically for company commanders, so that they could meet and discuss operations and lessons, and be retrained if necessary.[118] This decision was accepted, but was not the end of large-scale operations in 1950. This transition was not complete until late in 1951, when a fundamental shift from large- to small-scale operations occurred. Battalions began to deploy companies into the jungle to set up patrol bases.[119] Reconnaissance patrols were sent out to scout, if necessary supported by larger fighting patrols to destroy or ambush CT positions. By the time this transition became apparent, Briggs and others had accepted the fact that large-scale operations were not effectively destroying the enemy or denying access to villages for the CTs.[120]

The second major debate centred on whether there was a need for a new, all-encompassing jungle pamphlet for Malaya Command. The final decision was that the pamphlets that had been created at the FTC, along with various others produced elsewhere, were sufficient.[121] This last conclusion was a mistake but on the whole the objectives of the conference were to be commended. It was clearly an effort to centralise planning on tactical issues, and to focus effectively on the problems specific to a counter-insurgency campaign.

In the period from 1950 to 1952, Malaya Command made significant progress towards common doctrine and structured assessment of past mistakes by both senior and junior officers. FTC continued to train incoming officers and NCOs. Company commanders from units already stationed in Malaya were sent on two annual courses, and disseminated the information learned there to their officers. Battalions shifted their efforts from large-scale to company-sized operations, using smaller patrols. Some units held post-mortems on successes and failures in patrolling and ambushing. Such information was disseminated within the battalion and might pass back along the chain of command, to brigade level at times.[122]

Throughout the 1950s, the units of Malaya Command carried out a series of retraining exercises as full formations.[123] Since British Army units received an annual intake of National Service recruits and short-term commissioned officers, the need for battalions to retrain regularly outside the operational area became apparent. When new officers and recruits arrived to battalions in the field, some had already gone to the FTC; the others were trained by the officers and NCOs of a given company.[124] This structure was effective to a certain extent, although senior officers felt that battalions needed to be withdrawn from duty periodically and retrained as a unit.[125] It was decided that British battalions would be withdrawn from active operations every three years, while Gurkha and Malaya regiments would be withdrawn annually.[126]

During the height of the Emergency, 1949–54, the retraining structure encompassed issues specific to jungle warfare,[127] and used a framework similar to the one developed for Burma. Units progressed from basic weapons training and field craft to more complicated tactics. J.B. Oldfield's *The Green Howards in Malaya* described it as 'an opportunity to collect their wits and think ... to study success and find out its secrets and to profit by the lessons of those failures which every battalion in that theatre at one time or another experienced'.[128]

While battalions focused their efforts on more small-scale operations, assessed lessons and attempted to apply them in practice, political events in Malaya took a turn for the worse. In October 1951, the High Commissioner for Malaya, Sir Henry Gurney, was ambushed and killed by the MRLA. Not long after, General Briggs decided to step down as a result of illness. At the same time as these events were taking place, a change in government occurred in London. In the subsequent restructuring, the government decided to combine the offices of Director of Operations and High Commissioner, which was a drastic but needed step. General Sir Gerald Templer was appointed as the new High Commissioner with greater power than Briggs had had. He recognised the value of Briggs' initiatives, and was able to carry them forward with greater freedom and authority. He continued the resettlement process, and additionally tied the police, military and civilian mechanisms more closely together.

Templer also continued ongoing tactical initiatives and attempted further to streamline processes. Templer and his military staff continued to emphasise the need for assessment after Templer noticed that some battalions had more success than others in the jungle, and that success often correlated with an ongoing assessment process within different battalions. Templer also implemented initiatives of his own, particularly the creation of a manual for all troops in Malaya. He countered the findings of the Urquhart Committee of July 1950, and ordered the creation of a standard pamphlet for Malaya Command dealing with jungle warfare, known as *The Conduct of Anti-Terrorist Operations in Malaya (ATOM)*. He stated:

> I have been impressed by the wealth of jungle fighting experience available. ... At the same time, I have been disturbed by the fact that this great mass of detailed knowledge has not been properly collated or presented. ... [T]his vast store of knowledge must be pooled. ... [T]he job of the British Army out here is to kill or capture Communist Terrorists in Malaya. This book shows in clear and easily readable form the proven principles by which this can be done.[129]

Templer called on Lieutenant Colonel Walker to write the manual. Walker drew upon the pamphlets he had created at FTC, Indian Army manuals and his own experiences while in command of the 1/6th Gurkha Rifles. He finished the book in two weeks and it was available for distribution to the men and officers of Malaya Command by late 1952.[130] Walker's manual is larger than either *The Jungle Book* or the *Jungle Omnibus*, partly due to the fact that it includes material on the background to the conflict and overall strategy.[131] Templer, in accordance with the general mindset of Malaya Command at the time, invited criticisms and improvements from GOC Malaya, 'who [would] produce a revised edition in six months time'.[132] Eventually the manual was to be updated twice, once in 1954 and once in 1957. It is a tribute to Walker's efforts and abilities that both updates included only minor changes, mostly focusing on improvements in weapons, air supply, or helicopters.[133]

The end in sight: 1953–60

In 1952, the military effort in Malaya was expanded. From 1952 to 1960, deployments of troops arrived from Australia,[134] New Zealand, Fiji, East Africa and the Rhodesias.[135] The operational capabilities of the British, Malay and Gurkha battalions had increased steadily since the declaration of Emergency status. The publication of *ATOM* provided a clear and consistent doctrine from which recruits and veterans could draw. Rates of success increased with the operational shift to small patrols, and growing numbers of CTs units were denied access to populated areas. Larger fighting patrols were sent out after receiving information from reconnaissance patrols or intelligence from the Malaya Police.[136] The days of the large sweep had ended, but, as their predecessors in India Command and the 14th Army had done, the senior and junior officers of Malaya Command continued to assess their performance and tried to learn from it.

General Hugh Stockwell, GOC Malaya 1952–4, developed comprehensive reports on the state of training and tactics in Malaya Command during late 1953 and early 1954. Stockwell, too, recognised the need for proper jungle training. In one of his reports, he asserts that

> the training of the soldier must aim at producing a man whose field craft ability is better than his opponent. ... [T]his requirement must be considered early in the soldier's training, anything which tends to make him noisy in movement should be modified. His weapon handling must be slick and instructive but it must be quiet. He must be taught to move soundlessly.[137]

Stockwell also embraced the strategy of small unit patrols; in writing on the role of the military in 1954, he asserted that:

> the task of the army is to give security in the framework of defence, by patrolling into the jungle, searching out and harassing the bandit lanes and by specific operations against bandit organisation. ... [G]o after them and destroy them and prevent them taking any offensive measures against us or allowing their leaders to put over any doctrines. ... [T]hey [the Army] are to disrupt their means of access to the populations.[138]

Stockwell further emphasised the particular role of the junior officer in the Emergency situation. In discussing officer training, he commented that

> all operations must be conducted in cooperation with the civilian authorities and police. It is important therefore officers are taught the system of local government, the organisation and powers of police and the powers of the citizen, and any emergency regulations which have been made.[139]

Bolstered by such support and encouragement at higher levels of the command structure, assessment continued to fill an important role at lower levels. Brigadier Dennis Talbot, commander of the 99th Gurkha Brigade in 1955, disseminated the HQ Malaya Command Training instructions nos. 1 and 2 for his units. He amended the instructions with his own thoughts, specifically to call for assessment at all levels and documenting a training structure for troops after they came out of operations. This encompassed: (1) post-mortems on the operations just completed; (2) practice to implement lessons learned; (3) jungle craft training for junior leaders; and (4) practice of special techniques as laid down by commanding officers based upon instructions issued from brigade or Malaya Command HQ. He also recommended that, 'in addition to instructions issued by higher command, a careful study of the battalion contact reports and lessons learned issued periodically by this HQ should be studied by commanding officers'.[140]

Even with involvement in the Emergency winding down in 1957, emphasis on training in jungle tactics and doctrine continued unabated. Lieutenant General Sir Roger Bower took over as GOC Malaya Command in 1957. Before his tour of duty began, he was given a series of annual reviews to read, all of which made clear that lessons had been learned and stressed the continuing importance of specialised jungle warfare training, jungle warfare doctrine, and small unit operations. The reports provide an excellent summarisation of efforts undertaken in the previous nine years. For example:

> Details of operational techniques developed in Malaya and applicable on a sub-unit level are fully covered in the pamphlet *ATOM*. ... throughout the campaign operations of large numbers of troops designed to clear specific areas of jungle lacked success. ... to prepare troops ... their introduction for operations must be through a Battle School in theatre itself. As the Far East Training Centre tactical advanced parties have been trained by the Centre Staff and have been responsible for passing on this training to their units. ... good leadership and enthusiasm can produce that high standard of operational training and particularly of operational discipline without which any unit in the jungle is a liability.[141]

Conclusion

Neither the Indian Army of 1942–3 nor the British Army of 1948–9 was prepared for the problems it faced in undertaking combat in the jungles of Burma and Malaya. Both forces recognised their shortcomings, as well as the need for reform and common doctrine to reverse the previous defeats. The learning process for both theatres followed a similar path. Malaya Command in 1948, however, had a significant advantage once it re-discovered the lessons that had been learned in Burma and applied them, as appropriate, to its own situation.

Senior officers in both India and Malaya recognised the need for ongoing assessment of battlefield performance in order to identify changes in tactics and areas for improvement. The two armies faced two different kinds of enemy, but in both campaigns the jungle was still one of the greatest obstacles to overcome. Malaya Command drew upon the tactics developed for a conventional jungle war in Burma, and adapted them to an unconventional war in Malaya.

With the production of the *ATOM* in 1952, a documented link was established between the efforts of the Second World War and those of the Malayan Emergency. Troops earmarked for service in Burma and later in Malaya were required to undergo jungle warfare training, and in each campaign a formal doctrine was developed with which to train them. The *ATOM* built upon the foundation laid by *The Jungle Book* and the *Jungle Omnibus*, and like its predecessors was open to amendments and changes drawn from lessons learned on active service. Additionally, the *ATOM* did incorporate further developments in the thinking of the British Army. It included lessons and advice relevant to treatment of the civilian population, police cooperation and the political nature of the Emergency. The *ATOM* served as a basic introduction to counter-insurgency strategy, and its structure would serve as a template for armies around the world to follow for the next twenty-five years.

The second question that this book asks concerns whether participating in the Malayan Emergency made any contribution to the United Kingdom's capacity to fight a global war. This question is more difficult to answer, since this contingency has not yet arisen to be tested. In examining the British Army's performance in Malaya, and the British and Indian Armies in Burma before that, there are two significant developments with long-term impact on the British Army's capacity to fight wars, whether high- or low-intensity. These are the ability to evaluate past performance and apply lessons to future operations; and the growing focus on junior leadership skills. Both of these strategies have served the British Army well in the numerous low-intensity campaigns that have occupied its attention in the second half of the twentieth century, allowing it to rebound from initial errors and failures to achieve success generally.

Notes

1 *The Jungle Book*, 4th edn, 1943, preface, p. 2.
2 Ibid.
3 The Japanese were considered to be jungle 'supermen' after their victories of 1942 and 1943. While they were successful initially, ultimately the British and Indian Army units learned to overcome the jungle, and with it overcame the Japanese.
4 See Public Records Office (PRO), WO 32/5680–5683, Oriental and India Office Collection, British Library (OIOC, BL) Tactics and Training of the Imperial Japanese Army.

5 See L/WS/1/376 Equipment for Training Personnel of British Battalions in Burma; L/WS/1/267 Quarterly Returns of British Army units in Burma; L/WS/1/261 HQ Army in Burma Progress Reports, British Library.

6 30 November 1940, L/WS/1/261 OIOC, BL.

7 10 April 1940, L/WS/1/376 and WO 172/929 March 1942 Public Records Office (PRO), 1/11th Sikhs and Interview with Major T. Kirkwood, 15/3/2000.

8 There had been a series of official as well as personal thoughts regarding 'bush warfare' since the turn of the century. Some titles were: Major C.B. Morgan, *Hints on Bush Fighting*, London: Macmillan, 1899; Lieutenant Colonel T.O. Fitzgerald, *Bush Warfare: Notes from Lectures*, London: Macmillan, 1918; and Royal West African Frontier Force, *Notes on Training in Bush Warfare*, 1938. While this material was a good starting point for jungle lore and other points, many officers serving in 1941 or 1942 did not see this material. The material that was available was very basic when compared to the later editions of MTP no. 9 and the AITM.

9 See I. Stewart, *History of the Argyll and Sutherland Highlanders 2nd Battalion, Malayan Campaign, 1941-1942*, London: T. Nelson, 1947; A. Rose, *Who Dies Fighting*, London: Cape, 1944; and T. Moreman, *The Jungle, the Japanese and the British Commonwealth Armies at War, 1941-45: Fighting Methods, Doctrine and Training for Jungle Warfare*, London: Frank Cass, 2005.

10 6509-14 TS Copy War Diary 5/2nd Punjab Regiment, National Army Museum (NAM).

11 The 22nd Australian Brigade also attempted to develop tactics for fighting in the jungle. While it performed well at times, it was stationed in the southern province of Johore, and thus came into the fighting fairly late in the campaign. See L. Wigmore, *Australia in the War of 1939-1945 Series One Army: Volume IV The Japanese Thrust*, Canberra: Australian War Memorial, 1957.

12 Interviews with six surviving officers from the 1st Burma Campaign indicate that the main priority was survival.

13 The Army in India Training Memorandum (AITM) No. 14, February 1942, included observations of the Malayan campaign. One argument presented was that linear defence was impractical, and that it must be replaced by defence in depth. It also recommended the efficacy of re-supply from the air, and included observations from the 5/2nd Punjab Regiment on problems and solutions for operating in the jungle.

14 AITM, No. 15, p. 2.

15 AITM, No. 15, pp. 3-4.

16 The AITM and later *The Jungle Book* and *Jungle Omnibus* also included lessons learned by the Australians in Malaya and New Guinea and American observations of fighting the Japanese.

17 This school was later taken over by GHQ India and moved to Sevoke in 1943. A second jungle warfare school was also created at Shimoga in spring 1943. Correspondence with Dr Timothy Moreman.

18 The 23rd Indian Division, commanded by Major General Reginald Savory, was newly created. It was sent to the Imphal plain in Assam as the remnants of BURCORPS withdrew from Burma, and became involved in many of the subsequent tactical developments due to the fact that it was serving alongside the 17th Indian Division. The two divisions held joint exercises and Tactical Exercises Without Troops (TEWTS). Lieutenant General Reginald Savory Papers, TS Diary (July), 7603-93 NAM. The 34th Indian Division (later 20th Indian) was stationed on Ceylon in early 1942. Their task was to defend the island from a possible Japanese invasion. The unit war diary of the 14/13th FFRifles of the 100th Indian Brigade provides detailed insight into the tactical development of the division. Early in February 1942, the battalion received orders to begin to

learn jungle warfare. Compass courses were set up and patrol exercises created. Over the course of the next three months the battalion worked its way from individual jungle training to company and later battalion jungle warfare exercises. The battalion received lectures from Malaya veterans, including Lieutenant Colonel Stewart. The battalion exercises gradually were expanded to include brigade level and later divisional exercises by the summer of 1942. February–June 1942, WO 172/938 PRO. The officers of the battalion also noted that the MTPs and AITM were read, discussed and incorporated into the various exercises and drills. Interview with officers of the 14/13th FFRifles.

19 The 17th Light Division attempted to create a more efficient divisional structure during 1942 and early 1943. One of its innovations was the light division, which was to consist of two brigades. Its structure was accepted by GHQ India in June 1943, and in late 1944 with general standardisation, a third brigade was added. See L/WS/1/1310 Army in India: Indian Light Division, OIOC, BL.

20 4 June 1942, WO 172/475 PRO.

21 4 June 1942, WO 172/475 PRO.

22 As part of their instruction, all officers were required to read the Cameron Report, recent AITM, and Malaya Report. From June until the end of the year, the 17th Indian Division HQ created seven training reports that incorporated further information created in the MTPs, AITM and reports from patrols in the Imphal area June–December 1942, WO 172/475 PRO.

23 During late 1942 and early 1943, GHQ India and divisions began to create specialised battalions such as reconnaissance (half motorised and half mounted) to serve alongside divisions earmarked for Assam and Burma. There were further developments in the summer of 1943.

24 Two different forms were created, the high motorised and low motorised elements. The 'low' establishment formed the major basis for divisions in Assam and the Arakan.

25 A series of ad hoc formations had been created over the previous few years that included motorised as well as mule transport. The organisational tables were inconsistent, leaving some units entirely road bound and others mule based. This, in turn, affected the speed of operations. See L/WS/1/1333 Army in India Organisation of Animal and Mechanised Transport Divisions, OIOC, BL.

26 26 October 1942, L/WS/1/1333 Army in India: Organisation of Animal Transport and Mechanised Transport Divisions, OIOC, BL.

27 The 39th Indian Division was earmarked as well but took longer to begin the process.

28 18 January 1943, L/WS/1/1333 OIOC, BL.

29 The divisional training team was headed by Lieutenant Colonel Marindin, a Burma veteran.

30 21 April 1943, WO/172/2647 PRO.

31 Ranchi became the major staging area for all troops heading to the front in either Assam or the Arakan. Over time, training was stepped up in the area especially in jungle warfare as reinforcements arrived. The major divisions (such as the 7th, 5th and 20th) carried out further exercises in the area before heading to the front.

32 The 2/1st Punjab Regiment was part of the 14th Indian Division. They were given two jungle warfare demonstrations in December 1942. 4 and 8 December 1942, WO 172/903, PRO. In January, they were given four days to learn the basics of jungle warfare. 3 January 1943, WO 172/2367, PRO. This was clearly not enough, as their performance indicated, especially on 18 February. In an attack, the battalion failed to reconnoitre the area properly; when they launched a frontal attack, their men were cut to pieces. WO 172/2367, PRO and interviews with officers of the 2/1st Punjab. To make matters worse, reinforcements to the

battalion lacked training in all areas. Lieutenant Colonel A.W. Lother attempted to rectify the situation with a short course emphasising lessons learned, but timing constraints and the need for troops at the front cut this short. 19 February–5 March 1943, WO 172/2367 PRO.

33 The 14th Division was reinforced over the course of the campaign by six brigades.

34 14 January 1943, L/WS/1323 Operations in Burma: Ciphers, OIOC, BL.

35 See B. Fergusson, *Beyond the Chindwin*, London: Collins, 1945.

36 *The Jungle Book*, MTP No. 9 (4th edn), became the first major single pamphlet for all units within India Command. It specifically states on the inside front cover that all British NCOS and officers should have a copy, and that all Indian officers should have one as well. The Infantry Committee's Report also recommended the establishment of an HQ protection battalion due to the Japanese Army's demonstrated ability to infiltrate and attack HQs of brigades and divisions. An infantry battalion was awarded this designation. For the most part, specific companies were deployed to each of the three brigade HQs and one to the divisional HQ. 16 June 1943, L/WS/1/616 Army in India: Divisional Organisations, OIOC, BL.

37 The divisions trained Indian Army officers and men, while the brigade trained British Army officers and men. See L/WS/1/1371 Report of the Infantry Committee, India 1–14 June OIOC, BL.

38 After the end of 1943, men who were earmarked for service in Burma generally received specialised training, either at a training division or jungle warfare school. Some men received jungle training, only to be posted to Italy. Interviews with over eighty retired Indian and British officers.

39 After the 1st Arakan operation, the 14th Indian Division was earmarked to be the first training division. It was sent to Chindwara, since the 7th Indian Division had already carried out and created a jungle warfare training ground in that area. Later in the year the 39th Indian Division (the 26th had originally been designated but this was changed) was earmarked as the second training division and established in Saharanpur, northern India. Each infantry regiment of the Indian Army had a training battalion or a joint battalion set up within one of the training divisions. The British 52nd Brigade was set up to train British reinforcements heading to British units. L/WS/1/1364 Formations of the Training Divisions in India, OIOC, BL.

40 Major General A.C. Curtis (14th Indian Divisional Commander) Papers, Imperial War Museum (IWM).

41 Letter from Lt Alan Burnett, 15/10th Baluch Regiment, training battalion, 14th Indian Division.

42 The 19th and 25th Indian Divisions.

43 The 5th, 7th, 20th (34th), 23rd and 26th Indian Divisions.

44 The 17th Indian Division; the 39th had become a training division.

45 The fourth edition specifically states that it supersedes the third edition of August 1942.

46 See D. Marston, *Phoenix from the Ashes: The Indian Army in the Burma Campaign,* London: Praeger, 2003, ch. 4.

47 *The Jungle Book*, p. 19.

48 The box formations had their first formal employment in the Western Desert. The Indian and British Army units adopted the formation to jungle and open conditions in Burma.

49 See *The Jungle Book*; *Jungle Omnibus*, Lieutenant General Frank Messervy and Lieutenant General Gracey Papers, Liddell Hart Centre for Military Archives (LHCMA), King's College, London.

50 See *The Jungle Book* and *Jungle Omnibus*.

51 See *The Jungle Book*, pp. 10–11 and 'System of Attack', Messervy Papers, LHCMA.

52 See *The Jungle Book*, p. 3.

53 SEATIC Bulletin 9/7/1946 Observations of the War in Burma, Lieutenant Colonel Fujiwara, Lieutenant General Geoffrey Evans Papers, IWM.

54 Japanese Studies of WWII No. 89 Operations in Burma 43–44, Evans Papers, IWM.

55 This is not to say that the fighting was over. The 5th Indian and 11th East African divisions followed up the Japanese troops withdrawing to Burma along the Tiddim Road and Kabaw Valley.

56 The mission began in June 1943.

57 While India Command had recognised the need for specialised training and reinforcements for units in Burma by mid-1943, in late 1943 the War Office in London still did not think that the A and MT division structure was needed. As the Committee on Organisation for War Against Japan stated, the War Office directive 'takes no account of the necessity for a normal infantry division or an infantry division properly organised with a view to jungle warfare'. 16 December 1943, L/WS/1/650 Committee on Organisation for War Against Japan, OIOC, BL.

58 Major General J.S. Lethbridge Papers, 220 Military Mission Report, 2 Volumes, LHCMA.

59 Lethbridge took a reduction in rank to Brigadier to serve on the staff of 14th Army HQ.

60 By the middle of 1944 there were five different divisions: the high and low A and MT division, two different amphibious divisions and a light division.

61 Not all divisions that were deployed in the fighting in 1944 had one, counter to the Infantry Committee's Report.

62 The battalion would deploy one company to each brigade as required. Minutes of meeting, 26–27 May 1944, L/WS/1/650 OIOC, BL.

63 Two field regiments of 25-pounders, one mountain regiment of 3.7 inch howitzers and an anti-tank regiment of six-pounders was added. This development highlighted the possible 'open style' warfare that could possibly be encountered in central Burma. 26–27 May 1944, L/WS/650, OIOC, BL.

64 See AITM nos. 24 and 25.

65 See *Jungle Omnibus*, 1945.

66 My personal copy was given to me by Major Scott Gilmore, 4/8th Gurkhas, a veteran of the Arakan and Assam battles.

67 While jungle warfare was a central theme during retraining, all units discussed how to adapt tactics to the open plains of central Burma. When the time came, the basics of jungle warfare were used but adapted to the conditions. Boxes remained, but on a larger scale. Aggressive defensive tactics continued. Some units saw action in jungle conditions in the Arakan as well as the Pegu Yomas. See Marston, *Phoenix from the Ashes*, ch. 6.

68 WO 172/4980, WO 172/4866 1944 PRO and interviews with Colonel John Hill (2nd Royal Berkshire Regiment) Major R. Williams (8/12th Frontier Force Regiment) and officers of the 4/4th Gurkhas.

69 See Marston, *Phoenix from the Ashes*, ch. 6.

70 Major General Frank Messervy had long been a proponent of specialised training for jungle conditions. He served as a divisional commander in North Africa. He created a number of jungle warfare training instructions, drawn from many sources, for his division after he took over command in the late summer of 1943. Many of these training instructions can be found in his papers at the LHCMA.

71 See Messervy Papers, LHCMA and WO 172/4439 PRO.

72 September 1944 WO 172/4439 PRO.
73 WO 172/5029 PRO and Messervy Papers, LHCMA.
74 Interviews with five officers of the 4/8th Gurkhas. Also see P. Davis, *A Child at Arms*, London: Hutchinson, 1970.
75 WO 172/5029 PRO.
76 Interviews with officers.
77 British units were suffering from reinforcement problems due to the repatriation scheme.
78 'Lecture Given in New Zealand Regarding Malaya', Major General Dennis Talbot Papers, LHCMA.
79 There have been many books written on the success of the British counter-insurgency campaign in Malaya. Some of the best general accounts include Sir R. Thompson, *Defeating Communist Insurgency*, London: Chatto and Windus, 1966; R. Komer, *The Malayan Emergency in Retrospect: Organization of a Successful Counterinsurgency Effort*, Santa Barbara: Rand, 1972; R. Clutterbuck, *The Long Long War: The Emergency in Malaya, 1948–1960*, London: Cassell, 1967; J. Coates, *Suppressing Insurgency: An Analysis of the Malayan Emergency;* Boulder CO: Westview Press, 1992; J. Nagl, *Counterinsurgency Lessons from Malaya and Vietnam: Learning to Eat Soup with a Knife*, London: Praeger, 2002.
80 The military destruction of the Japanese forces in Burma was the primary objective of the campaign, thus the military effort was the central piece of strategy.
81 *ATOM*, Part I, ch. III.
82 *ATOM*, Part I, ch. III, section 8.
83 Foreword, Brigadier E.D. Smith, *East of Katmandu*, London: privately published, 1976.
84 Major I.S. Gibb from the 1st Seaforth Highlanders stated that in 1948 his battalion had no jungle warfare training and had only carried out Aid-to-Civil-Power exercises. He had received jungle warfare training in India during the Second World War. The battalion training in jungle warfare took place later in the year when officers and NCOs were sent on jungle training exercises. Major I.S. Gibb Papers, 86/3/1, IWM.
85 Four regiments of the Gurkha Rifles became part of the British Army, while six remained with the Indian Army. In 1950, General Boucher described the problems that the Gurkha Brigade faced at the beginning of the Emergency. He stated: 'with an average of 300 all ranks in each battalion, with new officers, a handful of untrained NCOs … a new system of administration, and no time to remedy any of these deficiencies, you were thrust straight into battle, in jungle of which the great majority had no previous experience. There was no time to train the new recruits and they went untrained into battalions.'(7th D.E.O. Gurkha Rifles, Digest of Service, 1948–1950. The Gurkha Museum, Winchester)
86 'Lecture Given in New Zealand regarding Malaya' specifically stated that in 1948, three British battalions were suffering from rundown caused by frequent releases. The Gurkha battalions were very much under-strength and waiting to be built up with new recruits. Major General Talbot Papers, Liddell Hart Centre, King's College.
87 Ferret Force was originally named Jungle Guerrilla Force. An order in August 1948 specifically stated their purpose, 'to locate and destroy insurgent elements who are taking cover in jungle country. … to drive such elements into open country when they can be dealt with the Regular Army and police units'. This unit was to be top secret. These original orders in the end highlight that this force was too small to deal with the CTs and the Regular Army needed to be trained in jungle warfare. L/WS/1/1498, IOIC, BL.

88 The training doctrine most likely came from the various AITM, *The Jungle Book*, *Jungle Omnibus* and MTP 51 and 52, since these were cited as material for the future Jungle Warfare School set up by Walker. See WO 268/116 Quarterly Historical Report FARELF Training Centre, 'Training Report April to September 1949'. PRO.

89 T. Pocock, *The Public and Private Campaigns of General Sir Walter Walker*, London: Collins, 1973, pp. 86–8.

90 In 1948 there were about 6,000 combat and nearly 3,000 supporting troops in Malaya and Singapore; by 1953 the numbers had risen to 22,000 combat and 10,000 supporting troops. See R. Sunderland, *Army Operations in Malaya*, Santa Barbara: Rand, 1964, p. 24.

91 The original location was Tampoi Barracks. It was later moved to Kota Tinggi.

92 WO 268/116 'Training Report from April to 30 1949'. PRO.

93 R. Gregorian, 'Jungle Bashing in Malaya: Towards a Formal Tactical Doctrine', in *Small Wars and Insurgencies*, London: Frank Cass, 1994, p. 347.

94 See Major I.S. Gibb Papers, 86/3/1, IWM; he specifically states that he was sent to the demonstration platoon at the jungle warfare school on p. 97.

95 The 1/2nd Gurkha Rifles did not engage in jungle warfare exercises until March of 1949. The course encompassed twelve days of training, run by a cadre of eighteen officers and men recently returned from FTC. At the same time the battalion was still getting up to strength with new recruits. Over the following months more men were sent on training exercises, and the battalion created post patrol assessments. In documentation included in the appendix to the battalion's war diary, one officer calls into question some of the tactical and shooting abilities of his men, and stresses the need for improvement. This process of reporting and assessment followed much in line with various appendices of war diaries from the Burma Campaign. WO 268/673 January–December 1949, 1/2nd Gurkhas, PRO.

96 Sunderland, *Army Operations in Malaya*, p. 45.

97 It is interesting to note that in the Second World War this period ranged from one to two months. One possible reason for the truncated structure may have been the perception that the MRLA, although good jungle fighters, were not of the same calibre as the Japanese, and that as a result more specialised training with artillery and tanks was unnecessary.

98 Sunderland, *Army Operations in Malaya*, pp. 46–9.

99 WO 268/116 September 1948–January 1949, PRO.

100 WO 268/116 September 1948–January 1949, PRO.

101 There was one major tactical difference between Burma and the Malayan Emergency in regards to defensive positions. In Burma, the troops dug all round defences with slit trenches, clearings and sometimes overhead protection. While the troops in the Malayan Emergency deployed for the evening in an all round defensive position, they did not dig slit trenches or create overhead protection. The CTs were not as tenacious as the Japanese in attacking a defensive position. Interview with Field Marshal Sir John Chapple.

102 Arthur Campbell, an officer serving with the 1st Suffolks, who had gone to FTC in 1949, specifically stated: 'I drove them (his men) hard during these three weeks. I had to for there was so little time and so much to learn.' A. Campbell, *Jungle Green*, London: Allen & Unwin, 1953, p. 14.

103 WO 268/116 September 1948–January 1949, PRO.

104 WO 268/116 April to September 1949, PRO.

105 WO 268/116 September to December 1949, PRO.

106 Some units such as the 1st Suffolks learned their trade well. See A. Campbell, *Jungle Green*, London: Allen & Unwin, 1953; and L. Spicer, *The Suffolks in Malaya*, Peterborough: Lawson Phelps Publishers, 1998.

107 Served as GOC 5th Indian Division.

108 See H. Miller, *Jungle War in Malaya*, London: Arthur Baker, 1972, pp. 70–3.

109 This strategy originated in the 1930s and became part of the formal doctrine, as embodied in the writings of Major General Sir Charles Gwynn, *Imperial Policing*, London: Macmillan, 1934, and the War Office pamphlets (*Notes on Imperial Policing*, 1934; *Duties in the Aid to the Civil Power*, 1937; and *Imperial Policing and Duties in the Aid of the Civil Power*, 1949). Malaya solidified the practice and set the example for matching police/Army cooperation. Many former Indian Army officers had experienced the collapse of the Indian Police in the Punjab in the run-up to independence and wished to avoid a lack of a formally structured 'war by committee'.

110 See *ATOM*, ch. 3, sections 2 and 3, for descriptions of State War Executive Committees (SWECs) and District War Executive Committees (DWECs).

111 See Coates, *Suppressing Insurgency*, pp. 149–63.

112 Major R.E.R. Robinson of the Devonshire Regiment wrote in 1950, 'without in any way wishing to overstress the point, it is a fact that most successful operations have been those planned and executed at the company level. Personal experience has tended to show that the larger the operation and the higher the level on which it is planned the less chance of success it has.'(*Army Quarterly* LX No. 1 April 1950, London: William Clowes and Sons, p. 80)

113 He would remain until June 1952.

114 Many battalion and brigade commanders were present. The commander of FTC, Lieutenant Colonel J.H. Law, was present as well.

115 WO 203/38 Tactics in Malaya. Notes of conference 11 July 1950, PRO.

116 WO 203/38 PRO.

117 From 1952 they were known as the 22nd SAS Regiment. They tended to be used as a strategic asset of Malaya Command.

118 Lieutenant Colonel Michael Calvert was one of the main supporters of small patrols. WO 203/38.

119 The 1/2nd Gurkha Rifles in 1950 clearly supported the use of small patrols. The CO ordered that companies be deployed to set up patrol bases. Small reconnaissance patrols covered the areas all around the base and attempted to locate CTs in the area. See 1/2nd Gurkha Rifles, Documents of Historical Interest, 1948–1954, 'Directive No. 2 All Company Commanders 1950'. Gurkha Museum, Winchester.

120 See Coates, *Suppressing Insurgency*, pp. 159–63.

121 WO 203/38.

122 See J.B. Oldfield, *Green Howards in Malaya*, Aldershot: Gale and Polden, 1953, pp. 68–9, WO 305/217 and 915 1st Suffolks, The 1/2nd GR had an appendix specifically called 'Lessons'. WO 305/248 1/2nd Gurkha Rifles, April 1951–March 1952 and April 1952–March 1953, PRO. See also 1/2nd Gurkha Rifles Documents of Interest, 1948–1954, Gurkha Museum, Winchester.

123 Senior and junior officers noted that soldiers' marksmanship was generally not up to the required standard. Often when patrols came into contact, the men missed the enemy as he quickly fled. The first shot therefore became very important.

124 WO 305/915 1st Suffolks, January 1952–March 1954, PRO.

125 Colonel G. Elliott of the 1st Queen's Royal West Kent Regiment reported in 1952 that the 'battalion was withdrawn for two month's retraining ... this was much needed as the turnover of officers and men had reached exceptional numbers. ... standard of training especially shooting had fallen ... it must be remembered that new drafts to receive training for 6 weeks before being allowed on operations.' (Colonel G. Elliott Papers, LHCMA)

126 General Dennis Talbot Papers, 'Report on the Military Situation by Lieutenant General Sir Hugh Stockwell, 15 October 1953', LHCMA.

127 As the threat decreased, many units trained using more conventional tactics to prepare for redeployment back to Europe. Interview with Field Marshal Sir John Chapple, 6 December, 2002.

128 Oldfield, *Green Howards in Malaya*, p. 43.

129 Preface, *ATOM*.

130 Gregorian, 'Jungle Bashing in Malaya', p. 350.

131 See *ATOM*.

132 Preface, *ATOM*.

133 Compare the first and third editions of *ATOM*.

134 Some Australian officers and men with jungle warfare experience had been posted to the FTC during the early days. They continued to serve on the staff of the FTC throughout the conflict. J. Grey, *A Military History of Australia*, Cambridge: Cambridge University Press, 1999, pp. 203–4.

135 Units carried out the same training and doctrine established by the FTC. As with British battalions, a cadre from each of the units was sent to FTC, Kota Tinggi for jungle warfare training. Men received the *ATOM* upon arrival in Malaya. Australian units also received initial jungle warfare training at the Jungle Warfare School at Canungra, Queensland. An officer of the 3rd Royal Australian Regiment noted that the training at Kota Tinggi was 'good stuff'. See C. Bannister, *An Inch of Bravery*, Canberra: Directorate of Army Public Affairs, 1995, pp. 15–45.

136 After 1955, due to the success of the security forces, patrols took on a further development. Due to the fact that many CT patrols were bumped quickly, specific battalion commanders deployed patrols that covered the role of recce patrols but also were strengthened with more men to carry out the role of a fighting patrol if needed. Conversations with Dr Christopher Pugsley and Field Marshal John Chapple. This was not official policy, as the FTC still continued to train men in both styles of patrols and the *ATOM*, third edition, 1958, clearly states a clear role for recce and fighting patrols. See chapter VII of *ATOM*. After the Malayan Emergency, Commonwealth units operating in Borneo and Vietnam followed the pre-1955 practice of distinguishing forms of recce and fighting patrols. Correspondence with Professor Robert O'Neill, Intelligence Officer with the 5th RAR in Vietnam, 1966–7. See R. O'Neill, *Vietnam Task*, Melbourne: Cassell, 1968. The 5th RAR made a further development in recce patrols. The commander, Lieutenant Colonel J.A. Warr, created a specialist recce platoon, whose function was larger-scale recce patrols and was an asset and under the command of the battalion HQ. See P. Hearn and R. Kearney, *Crossfire*, Sydney: New Holland, 2001.

137 7/4 Lieutenant General Sir Hugh Stockwell Papers, LHCMA.

138 7/9–13, 5 January 1954, Stockwell Papers, LHCMA.

139 7/4 Stockwell Papers, LHCMA.

140 'Training Instructions for 99 Gurkha Brigade', Operational Training, Talbot Papers, LHCMA.

141 Lieutenant General Sir Roger Bower, P123 'The Armed Forces' 27, IWM.

5 Aden to Northern Ireland, 1966–76

David Benest

Asymmetric or low intensity conflict was the norm for the British Army for nearly every year of the twentieth century, this despite recent assertions that it is a hallmark of the twenty-first. Whether described as small war, civil disturbance, emergency, internal security, revolt, insurrection, rebellion, insurgency or, simply, terrorism, matters very little. The defining theme is more or less the same – a challenge by the use of force, usually terror, against the authority of the state, colony, mandate or province. This embraces any situation in which regular uniformed British servicemen or women or police forces found themselves in confrontation with irregular armed groups (not distinguished by uniform) whose purpose was overtly political rather than criminal, even though acts of criminality were invariably used.

Where ruthless dictatorships faced these situations, the maximum use of force to remove the threat usually achieved the desired end state, at least in the short term. This was, essentially, the 'British way' throughout the nineteenth century. The dilemma for the British, and other democracies in the twentieth century, was how to achieve the same ends without stripping away the edifice of 'civilised' government, noting that British soldiers acted as private citizens in uniform and were accountable to the rule of law, whether civil or martial, in such circumstances, both within the United Kingdom and during all the counter-insurgencies fought abroad. The British approach was quite distinct from, say, those of France or the United States. This chapter addresses the tension between these conditions in the context of the transition from Aden to Northern Ireland.

Much has been written concerning the motives, organisation, techniques and technology used by insurgents and terrorists. T.E. Lawrence wrote of three essential dimensions – time, space and will.[1] At the strategic level the insurgent has time on his side and aims to convince his opponent that the war will last indefinitely: no government can as willingly accept an indefinite time scale. The insurgent needs a safe haven from where he can strike as and when he wishes: this may be the Arabian Desert as in T.E. Lawrence's day, the Republic of Ireland in the context of the IRA's campaign in Northern Ireland, the jungles of Malaya, or the inhospitable mountains of Afghanistan. His aim is to create the effect of his own invulnerability when compared to the immense vulnerability of the state. Perhaps most important of this trilogy is the will to succeed versus the will to resist. By definition, the insurgent's is essentially an offensive strategy, whereas the government under attack is placed on the strategic defensive. Governments need to consider the will of the people in support of their goals: the insurgent can impose his will on the people through terror.

It is also important to establish the boundaries of effectiveness in a terrorist campaign. As Charles Townshend relates, 'As a threat to the safety of the state, terrorism is implausible if not absurd; but as a challenge to the state's monopoly of force and the broader sense of public security, it is acutely effective'.[2] Townshend goes on to make the point that terrorism has to do far more with 'an assault on national pride and honour' than with posing an actual threat to sovereignty, which remains the provenance of conventional or nuclear forces alone. It is also useful to distinguish between what Ralph Peters refers to as 'practical' terrorism and 'apocalyptical' terrorism, the former more or less regarded as a rational actor fitting the Clausewitzian mould, the latter more in the realm of the religious fundamentalist. It is also true that, by their very nature, democracies are at a disadvantage in the face of terrorism. The terrorist has virtually total freedom of movement and of association, an abundance of targets, and can use the legal system to his own advantage. The terrorist occupies the grey area between peace and war, able to exploit the media to his own ends and always in a position to conduct 'a calculated assault on the culture of reasonableness'.[3] This conundrum is central to British thinking on counter-terrorism: how to respond to a truly Machiavellian attack, where the only response that will be effective will itself bring into question the moral authority of the state. Albert Camus identified this dilemma in the context of the French in Algeria when he concluded that, 'It is better to suffer certain injustices than to commit them.' The liberal state chooses not to respond in kind because of the fear of losing the 'moral high ground'.[4]

Each counter-insurgency campaign, naturally, has witnessed novel tactics, techniques and procedures, yet more generally the broad characteristics of the British response have been the same. Indeed, if 1916–18 can be seen as 'the birth of the modern' (to use Jonathan Bailey's term[5]) as regards conventional war, so too can the Boer War be seen as a watershed in the British approach to 'small wars', in that the British parliament and media were to take an increasingly close interest in their conduct. This is not to say that the many colonial counter-insurgency campaigns of the nineteenth century were exempt from public scrutiny, as Ian Hernon relates.[6] Rather, as the author acutely portrays, these were in many cases – Morant Bay, Jamaica for example – best forgotten, given the painful circumstances surrounding so many of them. This chapter will attempt to portray the sense of continuity in this field which led, finally, to the ongoing deployment in Northern Ireland.

The British approach to counter-insurgency has also been codified, both through various doctrine publications intended for reading by the armed forces and also through a variety of 'text books'. Charles Callwell's *Small Wars* of 1896, together with Charles Gwynn's *Imperial Policing*[7] of 1934 were the first. Sir Robert Thompson's *Defeating Communist Insurgency*, written in 1966, and Julian Paget's *Counter Insurgency Campaigning* of 1967 reflected the fixation with Malaya. But there has been surprisingly little written by practitioners of counter-terrorism since then. Most notable were Frank Kitson's *Low Intensity Operations*, his contribution to resolving the Northern Ireland conflict in 1971,[8] and his later memoir, *Bunch of Five*.[9]

The so called 'tenets' of counter-insurgency were variously stated as: that a clear political aim is required; that security forces must act within the law; that there must be an overall political plan involving all government agencies; that the strategy is to defeat political subversion rather than the insurgent *per se* (i.e. no 'military' solution'); and that bases needed to be secured. Coordination of intelligence, well understood tactics, techniques and procedures, clear rules of engagement and the need for a unified command at each level have also figured. These aphorisms could as easily be summed up in Clausewitz's dictum that war is but a continuation of politics by other means. They could also be taken as a given for any form of conflict, yet have achieved the hallmark of a 'special way' in warfare, misapplied or misunderstood by British forces prior to Malaya and certainly so by the USA during all its counter-insurgency campaigns from Vietnam onwards.

A constant theme of all British counter-terrorist campaigns was the assumption that forces of the crown must at all times operate within the common law. This has been a source of angst for practitioners, as Charles Townshend recounts: 'The difficulty of finding an appropriate legal response to internal emergency has been recurrent and acute for Britain, with its liberal-democratic political self image and its common law tradition.'[10] The omission in British law of any third way between a state of warfare, where the laws of armed conflict apply to all servicemen, and the assumption of peace, where the common law applies, remains a crucial conundrum, as demonstrated in the Vici and Vici v. Ministry of Defence case regarding the use of lethal force by British troops on the streets of Pristina in 1999, with similar cases arising in Iraq in 2003–4.

This is not to say that laws were not enacted to permit the prosecution of a counter-insurgency campaign. There have been many – witness: the Defence of the Realm Act (DORA) of 1914; the Emergency Powers Act (EPA) of 1920; the Restoration of Order in Ireland Act (ROIA) of August 1920; not to mention the raft of legislation of more recent years, such as: the Emergency Powers Act (EPA) of 1973; the Prevention of Terrorism Act (PTA) of 1974; and the Terrorism Act of 2000. There were also developments at the international level with the 1977 Additional Protocols to the Geneva Convention, aimed in part to impose a legal code on wars of national liberation (yet never applied in the context of the British war on terrorism, despite the international nature of the claims of IRA/Sinn Fein) and the 1977 European Convention for the Suppression of Terrorism. Less well known were the Acts of Indemnity required to absolve participants from their actions, perhaps a crucial precondition to any conclusion of the present 'peace process' in Northern Ireland.

The paradox of all counter-insurgencies has been that no matter how many times the principle that terrorism must be defeated 'within the rule of law' is invoked, this has never been achievable without substantial changes to the law so as to permit a level of coercion that can allow a successful conclusion to the campaign. The abrogation of habeas corpus and/or the introduction of internment without trial were the norm of nearly all campaigns and for good reason, because they worked. Of course, there was a price to be paid in terms of public opinion but, in a situation where the courts were rendered ineffective and government had ceased to exist in all but name, internment on a surgical basis has had its place in the lexicon of counter-terrorism. This, in the climate of a human rights culture, was and remains repugnant to many, especially lawyers. Perhaps it was because the middle class so rarely had to endure the inevitable privations of a terrorist dominated environment that firm measures of containment were viewed so critically. This was less so for those most likely to be affected. Or in the plain language of one citizen of Ardoyne in West Belfast in 1975 when the Provisional IRA was largely rendered ineffective by selective internment, 'Just keep those bastards off our backs!'[11]

The 'British approach' was the reflection of a deep antipathy to anything resembling military rule or martial law, a tradition that can be traced to the administration of Oliver Cromwell. Just as with legislation, the efficacy of civilian control was hardly ever examined and, all too frequently, 'the illusion of civil control was to be preserved at almost any cost, even if this meant worsening a crisis'.[12] This approach, of the legal liability of every soldier, had itself been called into question as early as 1837, when General Charles Napier famously remarked, 'Shall I be shot for my forbearance by a court martial, or hanged for my over zeal by a jury?' A Home Secretary put it in similar vein in 1867: 'The Military are entirely subordinate to the civil power, but the Military Officer in charge of a party is in sole command over his own force and disposes of it to the best of his judgement.'[13]

Another constant theme of counter-insurgencies was the strategy of 'winning' hearts and minds. This was generally well intended and in some circumstances can be said to have been successful, possibly in Malaya. However, there is very little evidence that 'converts' to the 'British way' were ever in any large numbers. The 1947 Internal Security Duties manual is also instructive in explaining why 'hearts and minds' were not easily won over, since measures to be considered included: 'punitive searches'; 'raids of a disturbing or alarming nature'; 'punitive police posts'; 'collective fines'; 'seizure of property'; 'demolition of houses'; 'taking of hostages'; and 'forced labour'.[14] Bluntly put, coercion was the reality – 'hearts and minds' the myth. The hearts and minds that really did matter were those of the domestic British electorate, for, without its support, no government was able to sustain a commitment that was both indefinite, on occasions deeply embarrassing, and costly to the taxpayer.

The British also retained a tradition of forming local, temporary and ad hoc 'third forces' or gendarmerie in circumstances where regular forces were inadequate for the task – as in Ireland in 1919–21 with the formation of the 'Black and Tans' and Auxiliaries, in Palestine with the Trans Jordan Frontier Force (TJFF), and in Northern Ireland in January 1970 with the Ulster Defence Regiment (UDR), later the Royal Irish Regiment. So too was it necessary to develop a covert capability, as in Orde Wingate's 'Special Night Squads' in Palestine, the Special Air Service in the Malayan campaign, Kitson's 'Counter Gangs' in Kenya and the Military Reaction Force in Northern Ireland. In the first instance this was a reflection of the manpower pressures placed upon the British Army during periods of peace. In the second, it was a realisation of the severe limitations in the utility of regular troops in countering irregular forces.

Counter-insurgency used to be an inherently national issue involving the sovereign power alone. This is no longer so, as witnessed in Afghanistan, the Democratic Republic of Congo and Iraq today. Rules of engagement have been shown to be distinctly different in these theatres shared by coalition forces, with a marked tendency for the British to allow the experience of Northern Ireland since *circa* 1976 to colour their thinking. Having stated at the outset that the distinction between insurgency and terrorism has been largely academic, a more refined view places counter-measures in two distinct categories, the former in the context of armed forces primacy, the latter in that of police primacy. This study examines how British armed forces made the transition from that of counter-insurgency to a counter-terrorist force over the period 1966–76.

The British Army had considerable experience of counter-insurgency by the time the Aden campaign commenced. Although Malaya has become synonymous with the conduct of a successful counter-insurgency campaign, its particular characteristics, especially geography and ethnicity, made it less relevant than might have been assumed for the conditions of Aden. In comparison, the second[15] campaign of the twentieth century on Cyprus, which was to last under four years from 1954 to 1959, did provide such a proving ground.[16]

Although EOKA never exceeded 200–300 in strength at any one time, it had almost universal support from the Greek populace, including the Orthodox Church, in pursuit of *Enosis* – union with Greece. The (Greek) police was easily demoralised and corrupted, necessitating the deployment of seventeen battalions and one armoured car regiment in four regions, each controlled by a brigade-sized headquarters.

This campaign saw the invaluable role of helicopters and the realisation, after much time and waste since the Boer War, that large-scale cordon and search operations were of dubious value compared to small deployments, a continuity that exists to this day in Northern Ireland. That said, it was necessary from time to time to utilise the road system, though not without great risk from mines and ambushes. The armoured car regiment deployed to the island was a critical asset: 'Its mobility, excellent communications and morale effect was most marked. Nevertheless, the vulnerability of armoured cars operating in built up areas or by night must not be overlooked: in both cases they are easily ambushed.'[17] The campaign involved the loss of 105 British soldiers killed, with a surprisingly high number (forty-nine) of fatal accidents through negligent discharge of firearms. A further 603 were wounded. The police lost 51 killed and 185 wounded, whilst there were 238 civilians killed and 288 wounded. Only ninety EOKA were confirmed as killed. EOKA was never destroyed, though neither did it achieve its aim of *Enosis*. The Turkish invasion of 1974 (rehearsed during Exercise Deep Furrow, a major NATO exercise, in September 1973) put paid to any such aspirations.

The Cyprus experience was to be applied the following decade during the Aden crisis of 1962–7, which of course chronologically foreshadowed the subsequent deployment to Northern Ireland.[18] The critical question is the extent to which the experience of Cyprus/Aden prepared the British armed forces for what was to follow in the United Kingdom.

The Aden campaign again saw British troops overstretched. Faced with the Egyptian and then Syrian backed National Liberation Front (NLF), they competed with the Front for the Liberation of Occupied Southern Yemen (FLOSY), whose militant arm, the Popular Organisation of the Revolutionary Forces of FLOSY (PORF), consisted of about six gangs, each of twenty terrorists operating in Aden state. Little Aden favoured NLF, whereas the docks area of Aden supported FLOSY. Both organisations specialised in close quarter assassinations using 'blindicide' rocket attacks and machine gun attacks on check points and observation posts (OPs), remote controlled mines,[19] mortars, jumping mine booby traps and, most frequently, grenades, with 2,041 attacks of various natures over the period September 1966 to September 1967, killing 33 and wounding 354. The Federal Regular Army (FRA), itself originally the Federal National Guard (FNG), had been infiltrated by both terrorist movements, as were the police, especially Special Branch, which had been rendered completely ineffective by terrorist attacks. The local population was thus unwilling to inform on terrorists, leaving the security forces with no intelligence upon which to base their operations. This was referred to in the Supplement to Middle East Command Newsletter Number 6:

> I hardly dare repeat the recurring theme on the need for intelligence. The lack has hamstrung our operations and caused us unnecessary casualties. I am sure the point is not lost to the Gulf, but it may be harder to push the requirement in Whitehall because intelligence costs money.[20]

The operation was seen as falling into five distinct phases, according to Brigadier R.C.P. Jeffries.[21] Phase 1 lasted from February to December 1966 and was typified by 'normal' internal security operations. Phase 2 from January to April 1967 saw the use of crowds by terrorists as well as the demise of police control. Phase 3 from May to September 1967 was a period of high intensity street fighting, described by Jeffries as 'If not exactly Limited War, there was now full-scale Armed Insurrection'. Phase 4 from September to November 1967 witnessed the ascendancy of NLF over FLOSY. Phase 5 in November 1967 was the British withdrawal in a situation of 'minimum force' but on the condition that 'Had there been opposition to the withdrawal, every single weapon, ship and aeroplane available would have been used as required'.

The principles of the pamphlet 'Keeping the Peace'[22] were relevant but 'A great deal of the detail, however, was not applicable to Aden'. Jeffries emphasised the importance of mass arrests and tear gas as a means of crowd dispersal, methods later to be adopted in Northern Ireland. 'Tremendous use … [was] made of armoured cars and armoured personnel carriers', provided that in the narrow overlooked streets, the requirement for mutual support with the infantry was never forgotten, a 'lesson' that was repeatedly to be ignored in Northern Ireland. Jeffries was equally critical of the application of 'lessons' from other theatres:

> Participation, especially by officers, in one or more IS operation in other parts of the world does NOT mean that they know all about things in the new theatre. A liberal dose of humility is essential, whilst making use of previous experience, to learn from the experience of others.

A system of alert states was in force, from 'green', through 'amber' to 'red'. A wire 'border' surrounded Aden, with only four entrances, which were also military checkpoints. Clandestine patrols were used in preference to an overt presence, which was on occasions too dangerous – a reminder that where the local population does not wish to comply with the forces of law and order, no amount of 'hearts and minds' will make much difference. This was itself a forerunner to the clandestine operations initiated by Kitson in Northern Ireland. Interrogation methods were also used which, when adopted in Northern Ireland, led to indictment by the European Court of Human Rights as 'torture'.

By April 1967 four battalions, an armoured car squadron and one Royal Marine company were fully committed, their main tasks being manning road blocks and check points, both foot and 'mobile' patrols, crowd dispersal, searches, the enforcement of curfews in Sheikh Othman and gun battles with terrorists. The carriage of arms was legal in Aden but a crowd of more than ten was illegal: there was thus no need for the niceties of the banner as a prelude to the use of force. The idea of single well aimed shots had already been replaced by the more effective 'burst of Browning through the nearest likely window'.[23] Nor was this conflict a matter of use of medium and light forces alone. Centurion main battle tanks and armoured personnel carriers (AFV 432) also played their part.

By the beginning of September 1967 the situation was so serious as to be referred to in the Middle East Command Final Newsletter[24] as 'intensive street fighting' rather than 'Internal Security'.[25] The events of the June Arab-Israeli War had had a marked impact, leading to the evacuation of the Jewish population from Crater, together with 9,700 service families, none of whom was attacked. The crisis culminated on 20 June with widespread mutiny and a wholesale onslaught on British troops wherever they might be encountered. After suffering 23 killed and 28 wounded that day, a decision to withdraw from Crater to allow for a period of 'cooling off' was taken, a precedent to be repeated in Northern Ireland in the early 1970s when parts of Belfast and Londonderry became 'no go' areas by daylight. The move was in part to avoid the risk that troops engaged in 'hearts and minds' 'up country' should become targets for attack – again a salutary reminder that a clear distinction needed to be made between the activities of enforcing a counter-insurgency campaign and the ability to engage peacefully with those as yet not committed. Crater was reoccupied peacefully on 3 July.

The intensity of PORF terrorist activity in Sheikh Othman and Al Mansoura was described as a 'pattern ... of incessant and often prolonged small arms attacks', mainly directed at static positions. References to 'incidents' were by now somewhat misleading as these might last up to two hours with large quantities of ammunition expended on both sides. There were 769 'incidents' in August alone as compared to a total of 480 for the whole of 1966.[26] The spearhead battalion was deployed from the UK, thus allowing 1 Para's tactical area of responsibility (TAOR) of both Sheikh Othman and Al Mansoura to be divided, with 1st Battalion, Lancashire Regiment taking over Al Mansoura. By 24 September both areas were handed over to the South Arabian Army. During the period 25 May to 23 September, 1 Para had coped with a variety of attacks including: 577 small arms; 144 grenades; 77 blindicide rockets; 202 mortars; 11 energa rockets; 5 mines; and 8 'others', totalling 1,024 attacks over a four month tour.

The Aden operation, like all the other British counter-insurgencies of the twentieth century, was inherently 'joint', with as many as seventy-eight aircraft in the Gulf area in May 1967. The vulnerability of air bases was a serious concern, with seventeen aircraft damaged in the six months from May and a further five in the following period, together with eleven attacks on Army aircraft. The Hunters of No. 43 Squadron (total strength eight) were used on four occasions in a close air support role 'up-country'. Firepower demonstrations were requested on seven occasions and three pre-planned strikes took place. A limited night strike capability was developed using Wessex helicopters to illuminate targets with flares. Troop lift operations were also a frequent occurrence up until July 1967 when troops were withdrawn from 'up-country'. Wessex helicopters proved invaluable, in the evacuation of casualties, in flying re-supply missions for the South Arabian Army, and in cooperation with ground patrols using especially fitted spot-lights, a capability readily adapted to Northern Ireland as 'nitesun'. Operation Hydraulic saw the deployment of thirteen Lightnings Mk 6 from the UK in June 1967, an interesting development in the context of the Six Day War. Royal Navy aircraft flew nearly 200 sorties from HMS *Hermes* and Victorious during the first three weeks of May 1967.

The Royal Navy was also to play its part. It intercepted *dhow* traffic suspected of carrying arms and reinforcements from the Northern Gulf to the South Arabian coast; it countered unrest at Mikalla on 15 May; it oper-ated the military port at Aden in the face of a civilian boycott; it maintained a significant presence in early June, including ten HM ships and four Royal Fleet Auxiliaries; and it formed Task Force 318 to cover the final with-drawal, with nine HM ships, one submarine, three landing ships logistics, five landing ships tank and twelve Royal Fleet Auxiliaries present in the outer harbour, a power projection exercise unparalleled since Suez and not to be repeated until 1982.

During May 1967 British forces were also undertaking civil famine relief operations in northeast Kenya. Thus when consideration is given to the nature of the high intensity conflict in Sheikh Othman and Al Mansoura, together with the 'hearts and minds' campaign up country, it becomes apparent that 'three block war' in today's parlance is a more fitting descrip-tion of operations in Aden than 'counter-insurgency' or 'internal security'.

The 'up-country situation' has been well documented[27] by 45 Commando Royal Marines. The enemy's aim was simply to drive the British out. The tasks of 45 Commando included the protection of Habilayn garrison and the conduct of mobile operations to provide a military presence throughout the area, prior to and during withdrawal. This was a combined arms opera-tion, typically including support from the Saladins of an armoured car squadron, an artillery troop equipped with a 105mm pack howitzer, a troop of engineers mounted in Saracens, 'soft' troop-carrying Stalwarts, 81mm mortars, general purpose machine guns (GPMGs) for prophylactic fire, two Special Air Service squadrons, armed (GPMG) troop lift, airborne command post or Casevac Scout helicopters, Sioux light helicopters for forward air controllers, RAF Wessex support helicopters for troop lift, Shackleton aircraft as flare droppers and RAF Hunters for close air support. Observation posts (OPs) and night ambushes were the preferred methods of operation, a tactic not dissimilar to that evolved by Orde Wingate in Palestine thirty years previously[28] and also developed from the study of operations in Radfan earlier that year. Response times were critical and a fifteen-minute crash-out regime instigated. Hard roads permitted harassing fire onto approach routes at last light. Enemy casualties amounted to twenty-four confirmed kills and sixteen wounded, with only two friendly forces wounded. The 'results of good training and imaginative use of heli-copters emerged as the match-winning factors'.[29] Rules of engagement were simply to bring the enemy dissidents into the open and destroy them. That said, enemy activity was equally well planned and during some attacks on the base, 'troops were pinned down in their sangars by very accurate small arms fire at close range and fire could not be brought to bear on the enemy until about 10 minutes after the beginning of the attack'.[30]

The withdrawal of troops from Aden involved a combination of light forces not dissimilar to those utilised at Suez, the Falklands and during Operation TELIC in 2003. The Hunters last saw action on 9 November in support of the South Arabian Army when they dispersed an attack by 300 irregulars of the National Liberation Army from Yemen on the SAA company position of Kirsh. The RAF's new Hercules C-130 also made its debut. British casualties amounted to 122 killed and 871 wounded, of whom 102 killed and 789 wounded were servicemen. So ended the last major deployment of British troops prior to that into Northern Ireland.

Any attempt to understand the situation in Northern Ireland up until 1976 would be incomplete without some mention of the previous Republican insurgencies in the British Isles.[31] The present campaign has long antecedents, achieving greatest prominence in the twentieth century with the Easter Rising of 1916 by some 1,200 members of the Irish Republican Brotherhood (IRB), which was supported by the Kaiser's Germany. The offer of international support was a constant theme of all the IRA insurgency campaigns, this despite the fiction that described the situation in Northern Ireland as a domestic issue for the United Kingdom alone to resolve. A mere fifteen rebels were executed after the rising had been ruthlessly suppressed by the British Army but, in the mythology of Irish Republicanism, the linkage of state execution to martyrdom had been established. The scale of the rising was relatively small, costing some 500 dead and 2,500 injured, with 2,000 volunteers interned. Yet the political impact was immense, with Sinn Fein winning 73 of the 105 Irish seats at the 1918 Westminster elections.

Although there had been sporadic outbreaks of violence throughout 1918, the death of two members of the Royal Irish Constabulary (RIC) on 21 January 1919 at Soloheadbeg, County Tipperary, saw the commencement of the War of Liberation. Between January 1919 and October 1920, 492 police 'barracks' were vacated and destroyed, 21 occupied barracks destroyed and 117 members of the RIC killed. The wholesale attack on the morale of this force had its immediate effect, with 2,000 resignations from the RIC by July 1921, together with a coordinated attack on the RIC's intelligence agency by Michael Collins on 20 November 1920. In the words of Sir Nevill Macready:

> This once magnificent body of men had deteriorated into a state of supine lethargy and had lost even the semblance of energy or initiative when a crisis demanded vigorous and resolute action. The immediate reason was not far to seek. If an officer of whatever rank took it upon

himself to enforce the law this, as often as not, would be disavowed by the authorities at Dublin.[32]

The IRA may have consisted of no more than 3,000 'ill equipped activists' but it was described by de Valera on 30 March 1921 as

> a recognised state force under civil control of elected representatives of the people, with an organisation and discipline imposed by those representatives. ... The Government is, therefore, responsible for the actions of this army. These are not acts of irresponsible individual groups therefore, nor is the IRA, as the enemy would have one believe, a praetorian guard. It is the national army of defence.[33]

A typical leaflet explaining the new political order ran as follows:

> We hereby proclaim the South Riding of Tipperary a military area with the following regulations. A policeman found within the said area will be deemed to have forfeited his life, the more notorious police being dealt with as far as possible first. Every person in the pay of England – magistrates, jurors etc, will be deemed to have forfeited his life. Civilians who give information to the police will be executed, shot or hanged.[34]

The campaign required 80,000 British troops at its zenith, as well as the introduction of internment without trial of up to 4,500 suspects. Yet even this number of troops was regarded as insufficient to bring decisive victory and it was estimated that a further 20,000 troops would be required, together with the imposition of 'security zones' to permit large-scale cordon and searches, all at a cost of £100 million per annum. This was simply unaffordable to post-First World War Britain, especially at a time when similar costs had forced the government to adopt the strategy of 'air control' in the mandate of Mesopotamia – an option not applicable to the 'bocage' topography of Ireland. The IRA was capable of mounting some thirty attacks each week during the spring of 1921, rising to fifty-five prior to the introduction of a truce on 11 July 1921, the signing of a treaty in December 1921 and its ratification by the Dail on 7 January 1922 by 64 to 57 votes. That the IRA was near military defeat is not disputed, but that it scored a stunning political victory cannot be denied either, a recurring theme of the 'Troubles'.

The Civil War of 1922–3 was the inevitable outcome of the alleged 'sell out' by Michael Collins to anything less than a 'true' republic, and this theme of traitorous compromise to something less than that envisaged in the Post Office declaration of 1916 has recurred subsequently. The declaration at the Four Courts led, by 22 June 1922, to its siege by Free State forces and a war which was to last until May 1923, costing between 800 and 4,000 lives. How or why these figures can be so vague remains unclear. The Free State was ruthless in its determination to eliminate the Republican threat. Special military courts were set up on 2 October 1922 and seventy-seven members of the IRA executed, three times the number put to death by the British in 1919–21. By 1923 the IRA was exhausted and it suspended operations in April, an enduring testimony to the reality that insurgency can be defeated by 'military' means if the will to do so is there.

This did not prevent the continuation of intense debate over ways and means. 'Abstentionism' remained a key issue for Republicans. For the IRA, the Convention of November 1925 saw the severance of ties between the IRA and Sinn Fein over de Valera's hint that he intended to enter the Dail. This was formalised at the Sinn Fein Ard Fheis of March 1926, when de Valera withdrew from Sinn Fein to form the Fianna Fail, entering parliament on 12 August 1927. It was not that de Valera did not believe in the use of force to liberate Ireland from the British. Rather it was its efficacy that was in question: 'There are times ... [when] because the odds are too enormous, or because the opportunity for swift blows for liberty do not exist, the use of force does not further the cause of national independence'.[35] Nor was there any doubt in Republican minds of the relationship between Britain and the Irish Republic that year, described quite simply as 'a state of war'. This was repeated by the Adjutant General of the IRA in his statement of 31 January 1933: 'Sovereignty and Unity of the Irish nation are inalienable and non-judicable, and the IRA cannot relinquish or surrender these fundamental principles, which are a sacred trust'. The IRA Council statement of 22 April 1933 was equally unambiguous: 'The Army is the leadership and the vanguard of the historic struggle for national freedom and for economic liberation'.

It should thus be hardly surprising that the IRA's campaign resumed in March 1936, leading to its proscription in Eire. Following the IRA Convention of April 1938 an ultimatum was served upon the British government to withdraw its forces from Northern Ireland, failing which a bombing campaign was to commence upon the British mainland, under the aegis of the well worn dictum, 'England's difficulty – Ireland's opportunity'. Under the 'S' Plan (after Seamus O'Donovan, former IRA Director of Chemicals), the IRA launched 291 attacks by the end of 1939, leaving seven dead and ninety-six injured. The campaign merely enraged British public opinion, providing the political climate for Parliament to pass the Prevention of Violence Act, which allowed for the deportation of 145 IRA suspects from mainland Britain and two executions after the Coventry bomb attack, in which five civilians were killed. Intelligence that the IRA was also engaged in intrigue with Nazi Germany allowed for the reintroduction of internment in January 1940 under the Emergency Powers Act, leading to the cessation of the campaign by the IRA in mid-1940.

The IRA was to remain dormant for the next sixteen years. That Northern Ireland should be the focus of its next campaign was first suggested in 1934 by Joseph McGarrity of Clan-na-Gael in the USA, who saw the six counties as the logical main effort for the IRA. A degree of campaign planning for this took place in the 1930s and again in the 1950s, to include raids on army barracks in northern England for arms and ammunition between 1951 and 1955. In the meantime, Sinn Fein won 152,310 votes in the Westminster elections of May 1955, having been re-activated as the IRA's political arm in 1948.

The border campaign commenced on 11 December 1956, at a time of international crisis for Britain, with over 80,000 troops committed to the Suez operation. It intentions were more akin to a Maoist strategy of 'liberation of the countryside', which may have made some sense in the context of South Armagh, South Down and parts of Fermanagh, but was hopelessly optimistic with regard to the Protestant 'heartland'. By July 1957 the Republic of Ireland (ROI) had responded with the introduction of internment, as was done on the north side of the border. The entire IRA leadership was out of action by the end of 1958. The number of attacks was reduced dramatically from 341 in 1957 to 26 in 1960. Sinn Fein lost heavily at the October 1959 elections, with a 50 per cent reduction on its record in 1955. The campaign was declared over on 26 February 1962.

The appointment of Cathal Goulding as the IRA's Chief of Staff in 1962 heralded a strategic reappraisal. The IRA was to adopt a national liberation front stance on a class basis against 'gombeeism' (i.e. Irish business interests) and the Committee for Revolutionary Action (CRA) was formed in 1965, with links to the British Communist Party and Moscow – a move strongly opposed by the diehard nationalists such as Sean McStiofan. Planning for the next campaign commenced in the mid-1960s, replacing the guerrilla insurgency style of warfare that had failed during the border campaign with 'terror tactics only'. Minor acts of violence took place, such as the shots fired at a British naval vessel in Waterford harbour in the summer of 1965 and the burning of buses in May 1968. A Campaign for Social Justice was also launched in 1964, evolving into the Northern Ireland Civil Rights Association (NICRA) by January 1967, with a Republican minority within it acting as 'stewards', especially figures such as Eamon McCann.

Thus by 1969 the IRA had already experienced four major campaigns and ceasefires. Five decades of discrimination by the devolved government of Stormont, itself a reaction to the continuous efforts by Catholic 'nationalists' to reunite the island of Ireland, provided the pretext for the next. The initial 'phase' was typified above all by civil insurrection and ethnic cleansing. Bogside was taken over by 1,000 Catholics in January 1969, with further disturbances across the province. An electricity sub-station was attacked on 30 March, leading to the mobilisation of the Ulster Special Constabulary to guard key points. Widespread disturbances followed in April, May, June and July. On 20 April alone, 209 members of the Royal Ulster Constabulary (RUC) were injured in Londonderry. Military assistance was provided for sixteen key points (KPs) in April. Severe riots broke out in the period 12–14 July in both Belfast and Londonderry. Troops were placed on standby to intervene if required by the RUC. Riots again broke out between 2 and 5 August in Belfast, and on 12 August riots spread around the province, killing ten, and injuring 1,600, including 800 RUC: 170 homes were destroyed in Belfast; sixteen factories gutted; and some £8 million in damage was inflicted on the Northern Ireland economy. Civil authority collapsed as the 3,000-strong RUC wilted under the strain of continuous rioting. Attempts by ROI ministers to meddle in the internal affairs of the United Kingdom by virtue of its constitutional claim to 'the north' led finally to a decision made on 19 August 1969 that the General Officer Commanding (GOC) should assume responsibility for security. The British Army was duly deployed to Londonderry and Belfast in a peacekeeping role, earning the gratitude of the Catholic population. Meanwhile, the RUC was to be disarmed and reformed under the Hunt Commission proposals.

The Army's order of battle prior to 1969 was that of a headquarters at Lisburn and 39 Infantry Brigade, co-located, with a battalion each at Holywood and Ballykinler, and a cavalry squadron at Omagh. A further brigade headquarters (24 Infantry Brigade) was deployed, along with eight battalions, and the Ulster Defence Regiment (seven battalions) was established on 1 January 1970.

Two years following Aden and ten since Cyprus, the middle to senior ranks of both the commissioned officers and non-commissioned officers of the British Army had experience of both campaigns, not to mention, in some cases, Malaya and Kenya. But were these experiences relevant to the situation now confronting the British Army in Ulster? Ostensibly the answer was 'yes' in that, as has been seen, the tactics, techniques and procedures that were developed in Cyprus and Aden were indeed employed in Northern Ireland in accordance with existing doctrine. Yet the unease with these as applied within the United Kingdom, as opposed to its colonies, was palpable. Also, the media, local civic and religious leaders, together with a culture of human rights, now dominated the politics of peacekeeping. The confusion was all too apparent, as some of the early reports and studies testify.

A study period was held by the GOC, Lieutenant General Sir Ian Freeland, on 5 December 1969.[36] The report made clear that the rioting in Northern Ireland was 'unlike that previously met elsewhere[37] ... The Army's previous experience, training and the techniques given in the pamphlets, do not fully cover this situation.' In these instances where riot and gun battles were interspersed, it was admitted that the Army 'must be prepared to fight' with 'one hand tied behind the back – using minimum force – against fellow citizens'.[38] As regards rules of engagement,

> There is a need for an intermediate weapon to bridge the gap between CS and the opening of effective small arms fire. It is for use when there is a militant minority trying, from different parts of the crowd, to cause escalation by throwing missiles etc. It gives the commander another option before fire has to be opened.[39]

His summary goes on to state, 'The degraded 7.62mm round would be of value. Its low velocity would reduce the chance of more than one person being injured, and the danger of ricochets would likewise be greatly reduced.'[40] 'The pressure on a commander facing a situation which may need a tough line, say, to open fire, is intense. Nevertheless, the correct decision must be taken irrespective of the presence of Press and TV.'[41]

Lieutenant Colonel Sibbald, Commanding Officer of 2nd Battalion, Light Infantry led the study on 'Crowd Control: The Problem of Escalation'. On the one hand, he made clear that fire was 'only returned by the Army at identified targets'.[42] Paragraph 18 went on to state: '*Target Acquisition.* There is the difficulty of identifying who gets shot? Should it be a petrol bomber? A crowd leader?' This was all in the context of operations taking place in darkness, where street lights 'are likely to be shot out as the riot progresses'.[43] Another commanding officer suggested that 'in order to fill the gap between the use of CS and firing for effect, a few rounds should be fired over the heads of the crowd. The consensus of opinion was against this.'[44] CO 2 LI

> said that firing over the heads of the crowd meant yet another step in escalation, and would be ineffective. He would prefer more resolute action earlier, and the opening of fire on the ringleaders to prevent the situation deteriorating into a full-bloodied street battle.[45]

It was also clear from the discussion that the colonial practice of a hand over point between control by the police and that of the Army was in operation. In such a situation, 'Clearly, the military commander must use his own judgement and exercise sole command'.[46] The GOC was all too aware of the sensitivities of this issue, since 'the whole question of how early the Army would be involved in rioting was now a political policy decision to be made in London'.[47] In one street 'battle' on the Shankill Road on 11–12 October, over 1,000 rounds were thought to have been fired, resulting in twenty-one wounded and one RUC constable killed. The tactical response to this was, to say the least, problematic, involving a transition from that of peacekeeping and deterrence to one of fire and movement. In a similar vein the study group agreed that 'petrol bombs thrown at troops could be taken as lethal weapons. However, proper warning must be given before opening fire.'[48]

Other non-lethal responses were considered, in particular the use of CS, which had been studied by the Himsworth Commission following the Londonderry riots of August 1969 and had 'proved generally favourable, so the use of CS has become politically acceptable'.[49] The capability of CS projection was under investigation at CDE Porton, as were the baton round, tranquilliser darts, electrified vehicles and dazzle devices, to name but a few. All this was in the context of finding an alternative to opening fire, a decision that seemed to be that of the commanding officer rather than of individual soldiers.[50] 'A weapon is needed which may be effective in dispersing the rioters, before finally having to open effective small arms fire.'[51]

The circumstances in which the order to open fire was given were made clear in the February 1970 brief, stating that single shots were to be used, that there would rarely be a representative of the civil authority present to consult before opening fire, and that a loud-hailer was more effective than banners or bugles, as recommended in the authoritative guide, 'Keeping the Peace', at paragraphs 32 to 42. The 'Yellow Card' guidance on opening fire was by then in place, having been issued on 25 September 1969.

The Commanding Officer of 1 Para, Lieutenant Colonel Mike Gray, led the discussion on Problem 4, 'Hearts and Minds'. He pointed out the many difficulties surrounding this, especially when soldiers were primarily required 'to deal forcefully with a riot situation, even prepared to shoot and kill'.[52] The Internal Security Brief for Units Due to Serve in Northern Ireland of 6 February 1970 is also instructive[53] in its emphasis on winning 'hearts and minds'. It seems that this was more a matter of wishful thinking than reality both in the Protestant 'heartlands', such as the Shankill, and certainly in all the strongly nationalist republican ghettos.

This first phase in the Northern Ireland campaign was to be short lived. The Provisional IRA (PIRA) was founded in December 1969 at the Army Convention with the formation of McStoifan's Provisional Army Council (PAC), formalised at the Sinn Fein Ard Fheis of 11 January 1970. The new IRA consisted of both Southern traditionalists (i.e. abstentionists) and Northern Republicans, mainly concerned with defending their own areas. Whilst OIRA saw the need for a political and military campaign, PIRA was focused on the latter. PIRA was highly active by 1971. That they had solid support was in Cathal Goulding's eyes quite clear: 'What helps the Provos most in the North was that every Catholic youth is a Provo at heart'. In the meantime, the furore in the Catholic community over the Official IRA's (OIRA) killing of Ranger William Best on 21 May 1972 was sufficient for OIRA to declare a ceasefire, though reserving the right to act defensively.

PIRA influence quickly spread through Belfast (nine PIRA units, four OIRA), Derry, North and South Armagh, South Down and Fermanagh. The PAC met in January 1970 to decide upon its strategy, seen simply as the defence of Catholic communities and an offensive against British rule, a policy made all the easier to foster and implement in the face of increasingly heavy-handed methods by the Army in Belfast, including the indiscriminate use of CS gas and baton charges. Barricaded areas to prevent Protestant incursions also doubled as 'no-go' areas, thus providing the safe houses from which PIRA needed to operate. In the early stages, the Army had become sucked into these areas by sectarian rioting, with PIRA actually acting to subdue the rioting youths for fear of a negative impact on its own image and recruiting and also of a premature clash with the Army. Other 'own goals' by the Army included its failure to defend the Short Strand in East Belfast from a Protestant 'invasion' in June 1970, which was repelled by PIRA which thus earned valuable support, to the detriment of the Army's own credibility. In the meantime, intelligence of large-scale arms smuggling into the Lower Falls Road led to the large cordon and search accompanied by curfew of 3–5 July 1970 – a turning point in the relationship between the Army and the Catholic community, not least because the action was illegal. Massive riots and a complete breakdown in civil order followed. Tactical success though it was, the Falls Road curfew was an operational and strategic political failure. Unlike Aden, the media brought every moment into the living rooms of residents in Britain, the European mainland and, most importantly, the USA.

PIRA's offensive began in October 1970, with 153 improvised explosive device (IED) attacks on commercial targets, most vehicle-borne, hence the term 'VBIED'. This was followed in January 1971 with PAC authorisation of attacks on the British Army itself, the first of which was the killing of Gunner Robert Curtis on 6 February 1971. That year saw a dramatic rise in the levels of violence, with 1,756 shootings, 1,515 IEDs and 174 deaths.

As GOC, Lieutenant General Ian Freeland was in no doubt that this was to be a long-term campaign.[54] First priority was given to bringing ringleaders of subversive elements to trial; the second to capture arms and explosives; the third to arrest rioters; the fourth to conduct protective and deterrent operations. There was to be no change to the responsibility under common law to use minimum force at all times, nor were there to be any radical changes in existing policies. Joint initiatives at lower levels were encouraged but large-scale searches were to be at the behest of the GOC. In conclusion, 'As the Prime Minister indicated last week, we develop our plans for a long haul and do not expect any single magic solution to our problems'.

The issue of rules of engagement became subject to an internal Ministry of Defence (MOD) working party, the Internal Security Working Party Report of 30 April 1971. That lethal force could and would be used as a means to disperse rioting mobs was not in doubt,[55] albeit as a last resort and only after all non-lethal means of persuasion had failed. These circumstances reflected the years of colonial internal security policy and practice, as witnessed, for example, by Major Bill Slim in India in 1923[56] and pursued up to and including Aden.

PIRA strategy required an end state whereby the British had withdrawn from Northern Ireland and a 32-counties republic formed. This was to be achieved through the engendering of a failure of political will on the mainland, together with the breaking of Unionist power by the imposition of direct rule, allowing the conflict to be portrayed as directly between PIRA, a national liberation movement, and the British Crown – an anti-colonial struggle which elsewhere (Palestine, Malaya, Kenya, Cyprus, Aden) had led to British withdrawal. In addition, it was intended to undermine the economy of the North, to make it ungovernable and to target the security forces (SF), agencies of government and commercial activities in a war of psychological attrition. Indeed, at one time, the number of British soldiers killed in Aden took on a special significance as a target against which campaign success might be measured. At the tactical level, the intention was to incite the British Army into Catholic areas, to create 'incidents' so as to undermine its credibility, use hooligans as cover and then mount attacks on the troops by gunfire. Civilians were, according to Marie McGuire, not deliberately targeted. Yet, since the creation of a state of widespread terror was part of the strategy, it was hardly surprising that civilian casualties rose alarmingly.

Reinforcements from the mainland included the deployment of brigade headquarters on an *ad hoc* basis, prior to the formation of Headquarters 3 Infantry Brigade in February 1972. For example, 16th Parachute Brigade Headquarters was deployed over the period 10 February–4 June 1971.[57] Based at Lurgan, the Brigade had responsibility for the security of the border as well as the towns around Belfast. Sealing of the border was 'quite impracticable with the forces available';[58] intelligence was poor; there was an urgent requirement to form 'bomb squads' to deal with the large number of IEDs; the report concluded prophetically that 'No solution will exist amongst a population whose moderates are not prepared to ostracise the extremist faction in their midst'.[59]

Such was the scale of PIRA-inspired terrorism that internment without trial was introduced on 9 August 1971. This was based upon intelligence on the Official IRA (OIRA), rather than PIRA and, of the initial 342 arrested, 116 were quickly released. The treatment of suspects was to be condemned by the European Commission for Human Rights as 'torture' and these practices ceased. Nevertheless, selective internment under Interim Custody Orders was to continue and 2,000 internees were held by the time of the next PIRA ceasefire.

Yet despite the increased levels of violence, the economic consequences were relatively minor – a projected growth rate of 10 per cent for Northern Ireland was reduced to 9 per cent in 1971. But inner city trade was badly hit, with a reduction of 30 per cent. Unemployment increased by 2 per cent (2.1 per cent on the mainland). The overall cost of the violence was on average £182 million per annum up until 1979.

More critically, a *Daily Mail* poll of September 1971 indicated that 60 per cent of the British public favoured the withdrawal of the British Army. The winning of 'hearts and minds' amongst the domestic electorate was not working and loss of public support on the mainland of Britain was a constant cause for concern, especially given the dramatic impact this had invoked in the USA in the context of Vietnam. That said, there was never equivalence between the loss of morale of British soldiers and that of the US Army at this time, at least for as long as the Army retained overall control, as in Aden.

By this time PIRA was sufficiently confident to initiate a five-point 'peace plan' which included the cessation of operations by the British Army, the abolition of Stormont, the release of political prisoners and compensation to 'victims of British violence'. If this were not accepted within four days, the violence would increase. This duly happened.

By now the instability generated by terrorism was sufficient to provoke the then Brigadier Frank Kitson, Commander 39 Airportable Brigade, responsible for security in Belfast, to write a paper outlining his concern at the lack of appropriate security guidance.[60] Kitson is quoted as saying, 'unless some general policy guidance is given on the long term situation our operations are very likely to seriously prejudice the future'.

In essence Kitson's paper[61] posited the overall policy of

> so weakening the IRA that a future political initiative can be launched under favourable circumstances. Despite the clumsiness of the Security Forces machine good progress was made ... largely because both wings of the IRA were also clumsy, and indeed much too big for the purpose for which they were designed to fulfil. It is likely that having fined down the enemy organisations to the extent that we have done, future successes will be increasingly hard to achieve from an operational point of view, unless we are able to make our own organisation very much more efficient. As you know we are taking steps to do this in terms of building up and developing the Military Reconnaissance Force (MRF) and we are also steadily improving the capability of Special Branch by setting up cells in each Division manned by MIO/FINCOs and by building up Special Branch's records with Int Corps Sections.[62]

Kitson saw the need for one of two options: the Belfast community was to be either 'integrated' or 'segregated'. The first of these would involve the breakdown of sectarian barriers, integrated housing estates and schools, and creating political allegiances based upon left versus right. Controlling bodies such as the Orange Order and the Republican Movement would need to be weakened. Security measures would include the divorce of Special Branch from the RUC, instead placing it under direct control of the security service.

In comparison, 'segregation' entailed the development of both communities separately but fairly, accepting the need for a Catholic police force to control the nationalist areas, which might entail, initially, the utilisation of vigilantes under British Army control. Legalisation of the Republican movement would also be necessary, which in turn could inflame the Protestant community into a full-scale onslaught against the British Army. Kitson concluded by emphasising that a delay in making practical decisions on these issues was unacceptable. 'It is no part of our business to recommend one solution rather than another but it is necessary for us to receive some direction beyond our immediate mission of destroying the IRA.'[63]

The scale of violence from July 1971 to early 1972 in Londonderry was such that 2,656 shots had been fired against the security forces, 456 nail or gelignite bombs thrown, and 225 devices had been exploded (mainly attacks on Protestant businesses), all leading to 840 rounds returned. In the two weeks prior to what is known as 'Bloody Sunday' on 30 January 1972, 319 shots had been fired, 84 nail bombs thrown, two soldiers killed, two wounded and £6 million in damages inflicted.[64] The tactic of using the cover of rioting hooligans for sniper attacks on security forces was well developed. The subsequent action by 1st Battalion, The Parachute Regiment (1 Para) was the subject of judicial inquiry by both Lord Chief Justice Widgery and the Saville Enquiry – thus further comment would be inappropriate at this moment.

All that can be said is that the severity of the situation was widely felt. For example, Lieutenant Colonel W.M.E. Hicks, Commanding Officer of 1st Battalion, The Coldstream Guards, in completing his post-tour report[65] was to remark (paragraph 19),

> A situation of virtually open warfare now exists in Northern Ireland. Under these circumstances the current rules for opening fire are considered to be too restrictive. Furthermore, the rules are in certain respects being broken almost daily and this undermines the authority of the Yellow Card [revised version received by his battalion in November 1971] and thoroughly confuses the soldiers (this in part appears to be over the practice of shooting out lights as a form of self-protection against snipers). In LONDONDERRY the need to be able to open fire on persistent rioters is urgent as the current baton round has far too short a range and CS has to be used in large quantities to even check the numerous and extremely hostile mobs encountered.

The value of armoured personnel carriers was also emphasised, 'They are essential for many operations' (paragraph 22), with a recommendation of as many as thirty-one to each battalion.

Hicks felt sufficiently strongly over the issue to write to both Commander 5 Airportable Brigade back in Aldershot[66] and Commander 8th Infantry Brigade.[67] In the first letter he referred to the inadequacies of the current rules for opening fire as 'too restrictive now that the IRA have resorted to what is virtually open war'.[68] In his view, soldiers should have the right to open fire in circumstances including: a person refusing to halt when called upon to do so; vehicles failing to stop at a road block or check point; any person throwing a petrol bomb after due warning; and[69] 'any rioter who persistently refuses to disperse when ordered – without warning'. He went on to state, 'If this (i.e. petrol bombs) had occurred in any colonial type situation I believe fire would have been opened against such persons on many occasions'[70] – a direct reflection of the status quo ante in Aden.

By July 1972 the future of Northern Ireland was under review at Prime Ministerial level.[71] The Northern Ireland Contingency Planning Report of 22 July dismissed any notion of a direct military assault on the Republican strongholds for fear of creating a state of permanent alienation. Withdrawal of troops was also ruled out. The third option, to remove arms and explosives through a massive reinforcement, would involve searches, interrogation and internment 'to administer a shock in the hope of forcing both factions to realise the necessity of an agreed political solution'. A redefinition of the border, together with enforced population transfers, was also considered, but it was concluded that 'We are extremely doubtful, therefore, whether the adoption of this policy would assist in providing a lasting solution to the problem of bringing about a stable society'. This was all predicated upon 'one specific course of action that Ministers might wish to consider if the security situation in NI had deteriorated so far that the Government were on the point of losing control of events', in what was described as a 'great emergency'.

The response would entail the deployment of every soldier in the United Kingdom or Germany who was needed for the execution of the task – some twenty units over eight days, effectively doubling the size of the force in the province from its current level, increasing to twenty-seven battalions in phase 2. This would entail eighteen of the twenty-seven battalions being deployed from Britain and nine from BAOR, leaving only two battalions in Germany.

Command was envisaged as running through the GOC, who would have 'full operational control of the RUC for security operations', giving him power of direction over the Chief Constable and giving the Director of Intelligence similar powers of direction over Special Branch, but not affecting routine police work. Control of the border, given its length and complexity, would initially involve a token force. Arrests were planned at a rate of 400 on the first day and 50 per day over the next two weeks, ensuring that 1,500 members of PIRA would be in custody, excluding the existing 300 internees.

As for rules of engagement, it was admitted that

> individual members of the security forces will be faced even more than usual with taking instant decisions on whether and how to act (e.g. when to open fire) and it is thought essential, if the security forces are not to be seriously inhibited, that such decisions should be taken without the fear of their legality being subsequently called into question.
>
> (Annex A, para. 17)

It was therefore considered 'crucial that … members of the security forces should be told of the Government's intention to seek an Act of Indemnity in due course for all actions taken in good faith by Crown servants in the course of the operation'. That said, acts of indiscipline would of course continue to be dealt with in the normal way. There would continue to be rules of guidance for when to open fire, amended to reflect the circumstances of the operation (para. 18).

Population transfers and/or a redrawing of the border were also considered. The inherent problems of doing either were considerable. A redrawing of the border west of the River Bann would place 238,000 Catholics and 227,000 Protestants within the Republic of Ireland. The transfer of Newry and retention of Protestant Londonderry would have left 130,000 Protestants in ROI. The fate of 114,000 Catholics in Belfast would remain as a conundrum. In all, population transfers would have entailed the movement of about 33 per cent of the entire population of Northern Ireland. Forced movement would have constituted a clear breach of Article 3(1) of the 4th Protocol of the European Convention. Any permanent barrier would have been in violation of the EEC treaty on the free movement of labour. Nor was it thought that such measures would in any sense lead to a cessation of the IRA's campaign – indeed, it would have been perceived as a 'step in the right direction' by Republicans. The report therefore concluded that 'It is extremely doubtful whether a transfer of territory, or population, could be effectively accomplished, or maintained, or indeed, even if it could be achieved, whether it would produce any worthwhile dividends'.[72] Internment on a large scale was considered as axiomatic to this contingency and was perceived as having a significant long-term effect, albeit with a severe political impact. All in all, the game was not going to be worth the candle. The Report therefore concluded that

> the Government's present policy of reconciliation, tempered with a firm but selective military response to terrorism, has more prospect of long-term success than any alternative being considered. Every effort should therefore continue to be made to ensure the success of the present policy; this may require some military reinforcements, a strict enforcement of the law, and possibly a greater local intensity of military operations.
>
> (para. 32)

The number of deaths that followed 'Bloody Sunday' (479 in 1972 alone) has rarely been taken into account when attempting to apportion relative accountability between the Security Forces and the IRA. The impact on PIRA was nevertheless sufficient for it to seek a negotiated ceasefire in July 1972 following the 72-hour truce of 10 March 1972. The ceasefire lasted for a very short period and by 9 July the campaign had resumed. The events of 'Bloody Friday', when twenty-one bombs exploded in Belfast, killing eleven and wounding 130, provided the impetus for Operation Motorman on 30 July, the removal of so-called 'no-go' areas in both Belfast and Londonderry by overwhelming force, at 30,000 in strength, the largest single deployment of British troops since Suez. This had the desired effect, forcing PIRA to withdraw to the border areas, with a concomitant drop in the number of IEDs from 180 to 73, shootings from 2,595 to 380 and soldiers killed from 18 to 11, altogether indicative of a serious loss of momentum by PIRA. According to McGuire, 'All the real IRA men are in jail now'.

Given the increased effectiveness of the security forces by the end of 1972, it should have come as no surprise that a mainland campaign was opened on 8 March 1973, timed to coincide with the border referendum that day, with three car bombs targeted at the Old Bailey, killing one person and injuring 147. Similar attacks took place in 1974 at the Tower of London in

July, a London restaurant in September, at Guildford in October and Birmingham in November, killing twenty-one civilians. They were checked by the passing of the Prevention of Terrorism Act (PTA).

The period 1971 to 1976 can, for want of a better term, best be referred to as a 'Colonial Strategy'. During these five years over one quarter of a million houses were searched, 5,000 vehicles checked every day, covert operations developed along similar lines as those used in Aden and an 'aggressive' attitude adopted towards the civil population generally. How this was meant to be in keeping with a policy of 'hearts and minds' was not all that clear. Nevertheless, with 2,000 of its members locked away, PIRA had been brought to its knees militarily and was forced to adopt its sixth cease-fire, which lasted from 22 December 1974 until 16 January 1975 and again from 10 February 1975 until March 1976. This was, typically, not accepted in the South Armagh/South Down area.

The expectation was of an end to PIRA hostilities in return for the phasing out of internment and a reduction of the Army presence in Catholic areas. Amongst the military the belief that PIRA was finished was commonly expressed. In part this was so, since the number of SF killed had dropped from 79 in 1973 to 50 in 1974, shootings were down from 5,018 to 3,206 and explosions down from 978 to 685. Also, the Gardiner Committee sat in January 1975 and recommended the end of both internment and special category status, thus partially meeting PIRA's demands, yet fatally undermining any hope of a longer-term reduction in support for PIRA in the second recommendation. Smith[73] claims that PIRA had itself 'forced' a cease-fire upon the British government. This seems somewhat far-fetched. Rather, the Social Democratic and Labour Party (SDLP) had managed to capture the 'hearts and minds' of the Catholic electorate with a 23.7 per cent count at the May 1975 Convention elections, which had been boycotted by PIRA. The arrest and trial of O'Connell in the Republic for IRA membership was indicative of how effective both governments had become. The PIRA/OIRA feud for 'ghetto supremacy', the formation of the breakaway Irish Republican Socialist Party (IRSP) and the Irish National Liberation Army (INLA), together with the security 'crackdown' of mid-1975 with over 400 charged, all pointed to a movement in the process of self-destruction.

The year 1975 also witnessed the 'Evelegh Study', probably the most detailed investigation ever into the legal relationship between the military and the law in the context of counter-terrorism.[74] This pointed to the shortcomings in the constitutional structure for suppressing civil disorder and terrorism (Ch. 1) and in the law controlling counter-insurgency operations (Ch. 2). It included proposals to rationalise the constitutional position (Ch. 3), and for a system of civilian control of a campaign such as Northern Ireland (Ch. 4). It stressed the requirement for dormant legislation (Ch. 5), for methods of surveillance (Ch. 6), for the infiltration of terrorist organisations (Ch. 7), and additional powers required for the military to make them more effective (Ch. 8).

As Commanding Officer of 3rd Battalion, Royal Green Jackets in the Upper Falls area of Belfast over two tours of duty in the late summer of 1972 and again in 1973, Evelegh had direct experience upon which to rely in producing his dissertation. His overall thesis was to query why 'the British Government's overall campaign to restore Ulster to a peacetime level of disorder was not more successful'[75] in seven years up until the time of his writing. He believed that success was in part denied through faults in the constitutional framework for controlling the campaign against insurrection and by shortcomings in the laws governing the operations of the security forces. 'The dilemma facing a democratic society is that the means needed to defeat terrorism and suppress insurrection are inevitably the ones needed to enforce a tyranny.'[76]

The paradox lay in the legal fiction that a soldier required to restore order in the face of insurrection had no greater powers to do so than any other citizen under common law, yet had an exceptional duty to act, similar to a constable or magistrate. In Evelegh's words,

> The fundamental problem is that the soldier when operating to suppress civil disorder is acting as part of a military force subject to military orders, indeed but for his status as a soldier he would not be there at all, but he is accountable to the civil courts under the terms of the legal fiction that sees him in these circumstances as simply a private citizen who happened to be passing the scene of disorder.[77]

There was at the same time no constitutional way in which the military could be put under the orders of the police or any other civil authority, although the military might be called upon by the magistrates to act, as defined in the Riot Act of 1714. The problem for the British Army in Northern Ireland was that the Riot Act had been abolished in 1967. 'Flexible law' was the outcome. The establishment of no-go areas, particularly in Londonderry, constituted a grave breach of law and the Army acquiesced in the defeat of the civil authority: 'The development of flexible law left the Army sloshing about on the troubled waters of Northern Ireland without the engine or rudder that should have been provided by constitutional certainty'.[78] Or again, 'What the troops lacked was not physical courage, or weapons, or will, but the legal confidence to do their duty and suppress disorder'.[79] It must be remembered that this was in the context of a typical four-month tour of duty where a battalion might suffer ten soldiers killed and forty-eight wounded, with up to 2,500 rounds fired against troops in a six-week period, with a ratio of deaths in South Armagh of 43 to 1 in favour of the IRA – in an operational area never actually typified as 'active service'.[80]

It was the removal of political status for prisoners, which came into effect in 1976, that was to provide the impetus for a revival of republican terrorism in what was described in the 1977 Staff Report as a 'long war' involving both armed struggle and a strong political dimension. The ravages of 1974–5 had also forced PIRA to reorganise into a cellular structure. Hence by 1981 Danny Morrison famously remarked: 'Who really believes we can win the war through the ballot box? But will anyone here object, if with a ballot paper in this hand and an Armalite?'[81]

Conclusion

The fixation with the 'Malayan' model of counter-insurgency has continued to dominate much debate concerning Northern Ireland. For example, writing in September 1993, Major N.J.R. Haddock, a student on the Army Staff Course, criticised the lack of a strategic objective so as to permit the military commander the ability to articulate a clear operational plan.[82] The absence of a Director of Operations was seen as a glaring omission, leading to a lack of cohesiveness and focus. This may well be true from time to time but it seems to this author that Haddock has missed the essential point, that there has been a long-term political aim embodied in the numerous agreements of the twentieth century, all of which saw re-unification as the ultimate goal, as and when the majority in the North wished it so. The compromise thus attained was the best that suited the particular conditions pertaining in Northern Ireland. To compare this with Malaya or any of the other counter-insurgencies, especially Aden, is simply irrelevant.

John Newsinger[83] seems to have come the closest to the truth in his remarks concerning the transition from a counter-insurgency strategy to that of a counter-terrorist strategy since 1976 under the policy of police primacy, despite the fact that in 1988 Margaret Thatcher very nearly reversed this process. The IRA was by 1994 incapable of operating in Londonderry, had a much-reduced effect in Belfast and only sustained itself at full capacity in South Armagh. Even here, the arrest of the South Armagh sniper team in March 1997 without a shot fired must be seen as the critical event. By then, of course, the IRA was well on its way to achieving its long-term goal, with increased support to Sinn Fein on both sides of the border and an increasingly despondent Protestant community in the North faced with the 'inevitability' of reunification. Just as in 1921, even though PIRA had in effect been militarily defeated, its political gains were immense.

The title of this paper was initially predicated upon Northern Ireland as 'the last frontier'. Events in Afghanistan and now Iraq have shown that this is far from true and that British forces will repeat the experiences of earlier campaigns of low intensity warfare, comparable with Cyprus, Aden, and Northern Ireland up until 1976. The events of 24 June 2003 in Majarr Al Kabir are worthy of recall, and not only just because this day witnessed the murder of six Royal Military Policemen in the local police station, which naturally became headline news. Rather, it was the infantry action which was occurring all that day in the same town that deserves mention. This was a high intensity engagement between an infantry platoon and up to 100 fedayeen, amply equipped with AK-47s, RPKs, 12.7mm heavy machine guns and RPG7s.[84] The account of the fighting could have come straight out of 1 Para's Aden experience, with stripped down Land Rovers quickly destroyed by RPGs, airborne reaction forces unable to function through their inherent vulnerability to the same form of attack, the similar vulnerability of medium forces such as Scimitar, and the general intensity of the firefights, requiring ammunition re-supply.

The main conclusions to be emphasised from the above account of the 'British way' in low intensity conflict are several. First and foremost, the idea that a 'template' can be applied to each and every situation has been shown to be a dangerous fallacy. Iraq in 2004 is not the same as Northern Ireland in the same year, though comparisons can be made between the two if we examine the early years in Ulster. Thus those who write doctrine for these circumstances are forever condemned either to produce such a bland guide to action as to be virtually meaningless – witness the frequent references to the irrelevance of 'Keeping the Peace' – or to recognise the unfolding of an inherently unique set of circumstances, in which case the dustbin beckons. Nothing could have made this plainer than the approach to Northern Ireland out of Aden. The 'template', if that is the most appropriate term, is rather that of the inherent danger of applying any counter-insurgency strategy which does not include an end-state that is achievable within a finite period. After over thirty years Northern Ireland remains the single highest concentration of British forces anywhere in the world.

A second observation or conclusion is that where, as in the British Army, a professional force becomes emasculated into little more than a gendarmerie, as in Northern Ireland since *circa* 1976, its ability to adapt to the realities of 'high intensity/low level warfare' is likely to become markedly weakened, a victory in every sense for the IRA's long war strategy. The siren voices that claim a pyrrhic victory of supreme courage when soldiers such as Corporals David Howes and Derek Woods were placed in the heat of battle on 19 March 1988, with scant arms and ammunition, and were left to die a most horrible death at the hands of a mob, might do well to consider how it was that a soldier's first duty, to engage with and if necessary to kill the enemy, could have become so confused and bemused by the constraints placed upon them. It is significant in this regard that the French-led deployment to the Democratic Republic of Congo in 2003 on Operation Artemis was regarded from the outset as 'robust' and concluded with a recommendation that basic fighting loads on the man be increased from two to three days expenditure! Evelegh's appeal of 1975 for the restoration of the Riot Act of 1714 seems all too apposite,[85] though completely unrealistic in political terms.

A third conclusion is that the events that actually dominated the British Army's thinking throughout the twentieth century and into the twenty-first have been shaped by this form of warfare, not preparation for high intensity operations. That this in some way enhances the ability of an army to change gear upwards is, to this author, a highly dubious proposition. It is agreed that our history of low intensity operations does allow for a much easier transition to these from high intensity, as was witnessed in Basra in 2003. The downside of the equation has been the British Army's reputation for a ponderous record in high intensity war fighting as a consequence of its almost continuous immersion in counter-insurgency.

The inherently 'joint' nature of low intensity operations has also been described and oft repeated throughout the twentieth century. Of course it is true that the Army has taken the brunt of the manpower required on each occasion but this is not to downplay the immense value of the Royal Air Force and Royal Navy on many occasions, including Aden, and the current deployments in Northern Ireland and Iraq. Where we are seeing a real change is not so much in this regard but in the multi-nationality of counter-terrorist campaigns, especially as witnessed in Afghanistan, Iraq and the Democratic Republic of Congo. This new phenomenon brings its own challenges, not least the viability of the rule of British common law governing our soldiers' actions compared to more robust rules of engagement applied by coalition partners.

In an era of the Strategic Defence Review's 'New Chapter' and its 'eight strategic effects' (prevent, stabilise, contain, deter, coerce, disrupt, defeat, destroy), of 'deliberate intervention', 'expeditionary deployments', 'focused intervention', 'power projection' and 'asymmetric warfare', to name but a few of the current buzzwords, all that can be safely assumed is that the likelihood of British forces coming into contact with irregular forces will be very high indeed. It would be a travesty if politicians were led to believe that in some miraculous manner, the 'British way' can achieve results without the necessity for a robust approach. The transition from Aden to Northern Ireland deserves study if only to remind decision makers of the harsh realities of counter-insurgency up to 1976. This must be compared with the subsequent years where, with the exception of a finely honed covert capability, the 'green' British Army was reduced to little more than that of a gendarmerie, ill fitted – despite assertions to the contrary – to the situations, such as in Iraq in 2004, in which it found itself fighting what is in reality a limited war.

Notes

1 Lawrence, T.E., *Seven Pillars of Wisdom*, Harmondsworth: Penguin, 1981, see pp. 193–202, 231–3, 345–8 for Lawrence's thoughts on insurgency .
2 Charles Townshend, *Terrorism: A Very Short Introduction*, Oxford: Oxford University Press, 2002, p. 116.
3 Townshend, *Terrorism*, p. 137.
4 Lieutenant General A.S.H. Irwin, 'The Ethics of Counter-terrorism – An Address to RUSI', *British Army Review*, Spring 2003, 131, pp. 6–10.
5 Jonathan Bailey, 'The First World War and the Birth of the Modern Style of Warfare', *Strategic and Combat Studies Institute Occasional Paper No. 22*, London: The Stationery Office, 1996.

6 Ian Hernon, *Britain's Forgotten Wars: Colonial Campaigns of the 19th Century*, Stroud: Sutton, 2003.
7 Charles Gwynn, *Imperial Policing*, London: Macmillan, 1939.
8 Frank Kitson, *Low Intensity Operations: Subversion, Insurgency and Peacekeeping*, London: Faber, 1971.
9 Frank Kitson, *Bunch of Five*, London: Faber, 1977.
10 Charles Townshend, *Britain's Civil Wars*, London: Faber, 1986, p. 9.
11 Related to the author whilst on patrol.
12 Townshend, *Britain's Civil Wars*, p. 15.
13 Ibid., p. 20.
14 Ibid., pp. 75–7.
15 See Gwynn, *Imperial Policing*, pp. 331–67 regarding the lesser known events of October 1931.
16 'Report on the Cyprus Emergency' by K.T. Darling, Director of Operations, 31 July 1959.
17 Ibid., para. 79.
18 Jonathon, Walker, *Aden Insurgency – The Savage War in South Arabia 1962–67*, Staplehurst: Spellmount, 2005.
19 See Maj. Allison, J.W. (retd), 'CDE's Support for IS and COIN Operations in Cyprus, Aden and Mauritius 1957–1970: Development of Fluorescent, Dye and Scent Marking Techniques', MOD RARDE Branch Working Paper 2/83(RIS), for an account of the attempt to counter the mine threat under Operation Kesgrave 1964–1966.
20 Ministry of Defence BR/1500/160CD(A) dated 19 June 1968, para. 22.
21 'Operations in Aden – Some Infantry Lessons', *The Infantryman* no. 84, November 1968.
22 Keeping The Peace (Duties in Support of the Civil Power) 1957, WO Code no. 9455, dated 10 April 1957.
23 See Report on Middle East Visit of Lt. Col. M.R. Johnston 14–25 April 1967, SCMF/Overseas/1, dated 3 May 1967.
24 Ministry of Defence BR/1500/160/CD(A), dated 19 June 1968.
25 Ibid., para. 23.
26 Ibid., para. 33.
27 Major A.G.H. Jukes, RM, 'A Report on Operations by 45 Commando Royal Marines in Area West 15 May to 26 Jun 1967', 45RM 7/11/101, dated 1 August 1967 SECRET.
28 For an excellent account of Wingate's role see Simon Anglim, 'Orde Wingate, The Iron Wall and Counter-terrorism in Palestine 1937–39', unpublished paper.
29 Jukes, 'A Report on Operations by 45 Commando', para. 25.
30 Ibid., Annex B, para. 14.
31 Richard English, *Armed Struggle – The History of the IRA*, Londong: Pan Macmillan, 2003.
32 Quoted in Tom Bowden, *The Breakdown of Public Security: The Case of Ireland, 1916–1921, and Palestine, 1936–1939*, London and Beverly Hills CA: Sage, 1977, p. 36.
33 M.L.R. Smith, *Fighting for Ireland: The Military Strategy of the Irish Republican Movement*, London: Routledge, 1997, p. 57.
34 *Hansard*, vol. 133, 25 October 1920, p. 1957.
35 *The Nation*, 14 January 1928.
36 Lt. Gen. Sir Ian, Freeland, 'Study Period Report', dated 5 Dec 69. Headquarters Northern Ireland.
37 Ibid., Summary, para. 1.
38 Ibid., para. 5.
39 Ibid., para. 12.

40 Ibid., para. 14.
41 Ibid., para. 17.
42 Ibid., Problem 1, para. 11j.
43 Ibid., Problem 1, Discussion para. 1.
44 Ibid., para. 4.
45 Ibid., para. 4.
46 Ibid., para. 13.
47 Ibid., para. 18.
48 Ibid., para. 28.
49 Ibid., problem 2, para. 2a(1).
50 Ibid., para. 5.
51 Ibid., para. 5.
52 Ibid., problem 4, para. 5b.
53 HQNI 51G, dated 6 Feb. 1970.
54 I.H. Freeland, 'Policy for Operations', NISEC 02, HQNI, dated 3 February 1971 paragraph 16, Headquarters Northern Ireland.
55 ISWP Report dated 30 April 71 para. 9c.
56 William Slim, *Unofficial History*, London: Cassell, 1959, pp. 73–99.
57 Major B.M. Tarleton, 'Report on Northern Ireland Tour', HQ 16 Para Bde 3006/1 OPS, dated 6 July 1971.
58 Ibid., para. 11.
59 Ibid., para. 36h.
60 See Howard Smith, Office of the United Kingdom Representative in Northern Ireland, letter to P.J. Woodfield CBE at the Home Office dated 30 December 1971.
61 Kitson, 4 December 1971, National Archives cj/3/98.
62 Ibid., para. 1.
63 Ibid., para. 12.
64 Report of the Tribunal appointed to inquire into the events on Sunday, 30 January 1972, which led to loss of life in connection with the procession in Londonderry on that day by The Rt. Hon. Lord Widgery, O.B.E.,T.D., London: HMSO, ordered by the House of Lords to be printed 18 April 1972, H.L. 101 H.C.220.
65 CO I COLDM GDS C2, dated 19 February 1972.
66 Ballykinler Camp BFPO 801 C2, dated 27 Oct. 1971.
67 FORT GEORGE 1CG Ops 34, dated 19 Nov. 1971 refers.
68 Ibid., para. 1.
69 Ibid., para. 3d.
70 Ibid., para. 5.
71 NI Contingency Planning Report, Cabinet Office 22 July 1972, National Archives, PREM 15/1010 TOP SECRET.
72 Ibid., para. 11.
73 Smith, *Fighting for Ireland*, p. 130.
74 'The Constitutional and Legal Problems of Involving the Military in the Suppression of Civil Disorder and Terrorism within the United Kingdom' by Lieutenant Colonel J.R.G.N. Evelegh MA, RGJ, Defence Fellow, Magdalen College, Oxford 1974–5, dated October 1975.
75 Ibid., p. 5.
76 Ibid., p. 91.
77 Ibid., p. 147.
78 Ibid., p. 37.
79 Ibid., p. 52.
80 Ibid., p. 54.
81 See J.M. Glover, 'Northern Ireland: Future Terrorist Trends', QMG Secretariat, D/DINI/2003 3/135, dated 2 November 1978.

82 N.J.R. Haddock, 'It Would Appear that the Principles Evolved from the Post-1945 British Experience of Counter-Insurgency have not been Applied Successfully in Northern Ireland at the Operational Level', ASC27 dated 10 September 1993.

83 John Newsinger, 'From Counter-Insurgency to Internal Security: Northern Ireland 1969–1992', *Small Wars and Insurgencies*, 1995, vol. 6, pp. 88–111.

84 Major C.P. Kemp, Para, 'Operation TELIC – Infantry Urban Close Combat', in *Infantryman*, Warminster: Headquarters Infantry, 2003, part 2, pp. 98–102.

85 Evelegh, 'The Constitutional and Legal Problems', pp. 238–40.

6　The unchanging lessons of battle

The British Army and the Falklands War, 1982

Simon Ball

A strange war

Short, victorious wars rarely teach profound lessons. On many levels this was true of the Falklands War fought in the spring of 1982. The Ministry of Defence's own report on the war opened with the words: 'The Falklands Campaign was in many respects unique. We must be cautious, therefore, in deciding which lessons of the Campaign are relevant.'[1]

The war was too short to allow for much in the way of adaptation or evolution. The Argentines invaded the islands on 2 April 1982. By 18 April the main outlines of a military plan to defeat them had been agreed.[2] 3 Commando Brigade landed at San Carlos Water on the west coast of East Falkland on the night of 20 May 1982. Eight days later one of its battalions, 2 Para, fought a battle for the settlements of Darwin and Goose Green in the south of the island. Advancing largely on foot, British forces were halfway to their main objective, Port Stanley, the capital of the island, situated on its east coast, by 30 May 1982.

The next day the second major element of British ground forces, 5 Infantry Brigade, came ashore. Elements of the brigade were taken forward by sea; during the course of this operation the RFA *Sir Galahad* was sunk by an Argentine air strike, which killed thirty-two soldiers from the 1st Battalion, Welsh Guards. The battle for the high ground overlooking Port Stanley began on 11 June, with an attack by 3 Commando Brigade. 5 Infantry Brigade launched a similar operation on 13 June. Argentine forces surrendered on the next day. The main land campaign had lasted for less than a month.[3]

The campaign was fought to one operational plan, with one team of commanders. It is doubtful whether, if either the plan or the commanders had failed, there would have been a chance to regroup and to try again. The land forces assembled in the first few days of the operation were those that fought the war: there was no rotation or reinforcement of troops. In addition to the celerity of operations, which affected all three services, there were additional factors that lessened the importance of the Falklands for the Army. Indeed, the participation of the Army in the conflict was curiously anonymous. One can explain this lack of impact in three ways.

In the year before the outbreak of the war defence had become a major public issue. The defence reforms launched by the Secretary of State for Defence, John Nott, had occasioned furious debate. Yet it was not the Army, but the Royal Navy, which was in the spotlight, as Nott tried to tie Britain ever closer to a NATO rather than a global role. It was the partial reversal of Nott's plans for the navy after the Falklands that attracted most attention.[4] The successes and failures of British naval power in the Falklands were minutely analysed. The naval post-mortem of the Falklands was long-running and international. It affected the debate about naval strategy within the Reagan administration in the United States, and it fuelled a vicious propaganda war between British and French defence contractors as to whose weapons systems had proved superior.[5]

The second factor that pushed the Army's experiences into the back-ground was the contingency of the units and personnel deployed. Although the Army provided the preponderance of ground forces used in the Falklands, the greater part of the Army's 'teeth' units was barely repre-sented. Neither line infantry nor armoured units played a significant part in the war. Infantry combat power in the Falklands was provided by specialist or elite units: the Parachute Regiment, the Guards, the Gurkhas and the SAS. It was these units rather than the Army as a whole that were seen to be involved. The battle of Goose Green was a paratrooper epic; the loss of the *Sir Galahad* at Bluff Cove was a Welsh Guards tragedy.

The initial assault force that landed in the Falklands and fought the first, dramatic, battles was 3 Commando Brigade. 3 Commando Brigade was a Royal Marine rather than an Army formation, albeit one significantly reinforced by army paratroopers. The land commander most in the public eye during the conflict was the officer who took charge of these landings, Brigadier Julian Thompson. Brigadier Thompson subsequently became an academic and a writer. He has analysed the ground war exhaustively in his writings, most notably in his best-selling history of the operation, *No Picnic*, published in 1985 and republished in 2001.[6] The naval commander in the South Atlantic, Rear-Admiral Sandy Woodward, also wrote a history of the war, *One Hundred Days*, which was published in 1992 and republished in 2003.[7] The barely disguised bitterness of the dispute between these two men about the direction of military operations during the war lies at the heart of an understanding of its lessons. Their first meeting on 16 April 1982 was little short of a disaster. Woodward regarded Thompson as his junior officer, and rated him lowly for bothering him with the detailed problems of the land operation.[8] Thompson believed he was Woodward's co-equal commander, and rated him lowly for his arrogance and ignorance of land warfare.[9] British problems with joint warfare were just as much a product of lack of communication within the navy, as of any failure of cooperation between the Army and the Royal Navy.

The other member of the command troika, Major-General Jeremy Moore, was also a Royal Marine. Moore commanded the later operations in the

Falklands with a full Army brigade, as well as the Army-reinforced Marine brigade, in action. He maintained a much more discreet public stance in the aftermath of the war, although, as we shall see, his position was in some ways the most compromised by the weaknesses of the command structure revealed by the post-war evaluations. The Falklands was Moore's final command and he retired to a career in banking, rather than continuing with his service career to the highest levels, like Woodward, or becoming a public commentator like Thompson. Thus, for many, the historical memory of the first stage of the land operation, carried out by Marines and paratroopers, remains more vivid.

Perhaps the most important factor affecting the Army's response to the Falklands, however, was its seeming irrelevance. The Army was undergoing a period of profound re-evaluation in the 1980s: but what it was re-evaluating was not the balance between continental and expeditionary wars but the nature of warfare on the North European plain. Under the leadership of Sir Nigel Bagnall, the commanding officer of 1(BR) Corps and subsequently NORTHAG between 1981 and 1985, the Army was embracing the concept of manoeuvre warfare.[10] This was more than a shift in tactical doctrine. Bagnall was slowly reintroducing the Army to a theory of victory. For over thirty years military planners had assumed that deterrence would prevent a war in Europe but that, if the unthinkable happened, NATO had no chance of winning a conventional war against the Soviet Union. Such a war would descend into a nuclear conflict within days, if not hours. The suggestion that this was not the case was immensely exciting for army officers of all ranks. The pros and cons of the manoeuvrist school engaged the attention of the Army's brightest leaders. The pages of the *RUSI Journal*, the *Army Quarterly* and the *British Army Review* were filled with their arguments. Despite the urging of editors, the Falklands War provoked no equivalent debates.[11] There had always been something of a division between the 'heavy' army in northern Europe and the 'light' army built around special operations on a global stage. Sir Peter de la Billière, the senior British commander during the first Gulf War, famously declared: 'I am *not* a BAOR man myself.'[12] In the 1980s the 'heavy' army and its concerns were in the ascendancy.[13]

'War makes rattling good history,' said Thomas Hardy, 'but peace is poor reading.' In the years after 1982, the Falklands was often used to provide dramatic illustrations in discussions of military matters across a wide range of issues ranging from high command to minor tactics. It is debatable whether it changed fundamental views on any of them. As an illustration of how the Army dealt with the lessons learned, the rest of this essay concentrates on a debate from each end of the spectrum; first the provision of infantry firepower, then the command and control of joint operations.

Firepower

There were a number of reports on the 'lessons' of the Falklands drawn up in the immediate post-war period. By far the most attention was focused on the 1983 Franks Report concerning the origins of the war. Lord Franks and his team exonerated Mrs Thatcher's government from any significant culpability for the war, ensuring that their report became a focus for political and public discussion. The main military issue addressed in the Franks Report was that of naval deterrence.[14] It thus had little resonance for the Army. The Ministry of Defence's own report on the operation, published in December 1982, was more important to the armed forces. Like the Franks Report, *The Falklands Campaign: The Lessons* was a carefully phrased document. Published only a few months before a general election in which the war was a major issue, it skirted around the controversial issues. Its main focus was on equipment rather than on the conduct of the campaign. Once more the debate about the performance of British warships took pride of place. No such debate, with political and budgetary aspects, applied to the Army. The report concluded that the 'present types of weapons proved effective but the infantry need to be supported by greater direct and indirect firepower in attack'. Direct firepower would come from Milan anti-tank guided weapons (ATGWs) and disposable light anti-tank weapons (LAWs). The Army decided to re-introduce 51mm mortars at platoon level. The only new type of equipment that was required was 'an area attack weapon such as a grenade launcher'.[15]

The sections on the military campaign were short and not particularly informative. The report argued that 'the most decisive factors in the land war were the high state of individual training and fitness of the land forces, together with the leadership and initiative displayed by junior officers and NCOs'. The most striking tactical lesson was that 'night operations and aggressive patrolling' were 'particularly decisive' for assaults 'conducted against a prepared enemy with clear fields of fire'.[16] The bland nature of the report led some commentators to believe that it was a bowdlerised version of a much more hard-hitting review carried out for internal military consumption. The Ministry of Defence has consistently denied that this was the case.[17]

This is not to say that a number of detailed operational assessments were not prepared. These assessments were coordinated by the Deputy Chief of the Defence Staff. Shortly after the Argentine surrender, for instance, a seven-man scientific evaluation team arrived in the islands to conduct post-combat interviews. They visited twenty-two Army units in the course of their investigations.[18] Apart from a few journalists, there were no observers present to offer an independent opinion. Major-General Edward Fursdon, a retired officer who had just published a well regarded history of the European Defence Community, had requested permission to accompany the Task Force with the view of writing an 'instant' official history of the war, but his offer was declined.[19] Fursdon had to content himself with writing a book about the post-war Falklands.[20]

Apart from assessing its own performance, the most obvious alternative source of information and opinion for the Army was from the 'other side of the hill'. The Army had a large number of Argentine POWs available as a potential intelligence resource. On 27 May 1982, HQ Land Forces, Falkland Islands was told by the MoD that, 'they could … authorise more detailed questioning of POW'. The British regarded the Argentines as having put up a lamentable performance during the campaign, however. The most memorable aspect of Argentine military practice for most soldiers was their habit of defecating in their own trenches, rendering capture of an enemy position a mixed blessing.[21] They thus had little interest in the views of the vanquished. Argentine sources were dismissed as being of 'marginal usefulness'.[22] The Army part of the post-Falklands review tended to be inward-looking, concentrating on the experience of its own units. Thus the Army view of lessons to be learned emerged almost entirely from within the Army itself. This lack of alternative sources of information or evaluation affected the way that the Army implemented changes, choosing to adapt the experience of combat to already formulated preferences, rather than *vice-versa*.

The generally applicable lesson that the maximum possible firepower be applied at the decisive point, and at that decisive point could be found and exploited best at night or in low light, was obviously correct. The purchase of 'starlight' night vision aids was accelerated. The use of firepower in a real war meant the expenditure of ammunition far beyond anything that peacetime planning envisaged. 'Quite clearly,' concluded one review, 'the limited war scaling of ammunition is unrealistically low and this could have proved disastrous had the land battle continued even for a short period of time.'[23]

This general underestimation of how intensely firepower would be used in any real war was compounded by the specific circumstances that arose during the campaign. The invasion force was supposed to achieve offensive mobility through the use of helicopters rather than by ground vehicles. Unfortunately, most of the force's heavy lift Chinook helicopters were lost at sea when the *Atlantic Conveyor* was sunk by an Argentine air attack on 25 May 1982. For much of the campaign the invading force had to operate with but one Chinook. Although 5 Brigade made some use of sea-lift, most ground troops made their way from west to east across the island on foot. An added premium thus attached to the lightness and manageability of equipment.

British troops in the Falklands were armed with two main infantry firearms. Both had been designed by the Belgian firm of Fabrique Nationale on the basis of lessons learned during the Second World War. The FN MAG (*Mitrailleuse d'Appui Général*), or General Purpose Machine Gun (GPMG), could be fired from either a tripod or a bipod. By general consent it was one of the finest weapons in its class, providing a high rate of fire with exceptional reliability. The SLR (self-loading rifle) was a semi-automatic version of the FN FAL (*Fusil Automatique Légère* – the Argentines used the automatic variant). Once derided as the 'mechanical musket' for its lack of long-range accuracy, this weapon too was prized for its ruggedness. Both machine gun and rifle fired the standard NATO 7.62mm round. The

Falklands review suggested that both these long-serving weapons were too heavy for general issue to the infantry, but that 'these problems will be eased considerably by the new small arms which should enter service in the mid-1980s'.[24] Table 6.1 compares the weight of the SLR and the SA80.[25]

It had been a long-standing ambition of the Army to introduce a small calibre rifle with high accuracy and low recoil into general service. The adoption of a revolutionary bullpup design – the EM2 – had only been narrowly thwarted by the Americans at the beginning of the 1950s. American adoption of a 5.56mm calibre rifle in the 1960s led to the project being resurrected in 1973. A decision in principle to re-equip the Army with the 5.56mm bullpup rifle had been taken in 1979. There was nothing inherently objectionable about the SA80 as it became known. Bullpups offered the advantage of a long barrel, and hence greater accuracy, for any given length of rifle. France and Austria introduced outstanding weapons in this format at the beginning of the 1980s.

What was more questionable was the decision to relegate the FN MAG to the sustained fire role, 'in the "golf bag" of weapons in platoon HQ and pintle-mounted on vehicles', and to replace it with a bipod-mounted heavy-barrelled version of the SA80 called the Light Support Weapon (LSW). The argument put forward for this change was the reluctance of infantry battalions to adopt a section organisation based around two four-man fire teams, as advocated by the School of Infantry, because each section only carried one GPMG.

Yet in the Falklands this problem had been overcome by the simple expedient of issuing more FN MAGs. As a result, 'many adopted the two fire team concept with great success'. Wartime experience did not, however, correspond with emerging tactical doctrine, and was therefore discounted. It was objected that in the Falklands 'the GPMG still remained the work horse of the section; taking on the major share of the firing, with the SLRs often doing little except provide protection for the guns'. What was supposed to happen in the new system was quite different. A section equipped with two LSWs would no longer have to rely 'solely on one GPMG'. Instead the section could engage the enemy with both its LSWs and its fully automatic SA80s. 'This should', it was argued, 'ensure that fire support is reliable and that there are no ominous silences during an attack while a GPMG gunner reloads or tries frantically to remedy a stoppage.'

The main argument against the standard issue of more GPMGs was their weight, and the extra complexity and weight that would be entailed by

Table 6.1 Weight of British service rifles in kg

	SLR	*SA80*
Weapon and 1 full magazine	5.06	4.98
Weapon and ancillaries and combat ammunition (120 rounds)	9.51	7.3

carrying two types of ammunition – 7.62mm and 5.56mm. The Falklands, with its logistical problems and enforced use of route marches, lent some credence to this argument. Yet the main role for the SA80/LSW combination was actually to be in armoured warfare when the infantry would be mounted in armoured personnel carriers. 'At close range, the volume of automatic fire from the section's weapons will be awesome', went the official version; 'this will be of particular importance in mechanised warfare when the enemy can be expected to debus either on or very close to the objective and there will be very little time in which he can be engaged'.[26]

The Falklands had reinforced the lesson of all modern conventional wars that firepower wins the infantry battle. It had also confirmed that the unavoidable pattern, for both offensive and defensive firefights, was dictated by the weight of fire from a machine gun. That experience was only partially heeded, however. It was overridden by a long-standing enthusiasm for personal marksmanship with the rifle. On paper the results of trials were impressive. Every soldier tested was able to pass marksmanship tests with the SA80, whereas only 72 per cent were able to do so with the SLR. The SA80's Achilles' heel proved to be its unreliability. Later trials, conducted when its operational shortcomings had become apparent, concluded that its success rate in carrying out its 'battlefield mission' was only 67 per cent.[27]

The post-Falklands enthusiasm for the LSW is much harder to understand. At the time machine gun experts pointed out that infantry firepower must be 'heavy, controlled and above all sustained to be effective'. In order to achieve sustained fire a machine gun must be belt-fed and have an effective barrel changing system to combat the deleterious effects of overheating.[28] The LSW was magazine-fed and did not have a quick barrel-change facility. It was thus suited only for single shots and intermittent bursts. 'Let us face it', a retired officer wrote, 'weapons of the automatic rifle type … are largely discredited as effective arms for the large-scale warfare, however useful they may be in the brief clash on a jungle track.'[29] Later tests showed that the LSW was capable of achieving its battlefield mission only 5 per cent of the time.[30]

In 1984 the Army put the LSW through a competitive trial with potential alternatives, ranging from the venerable Bren gun to FN's new Minimi 5.56mm LMG. The LSW was declared the preferred weapon and was introduced into service in 1986. A parliamentary enquiry later pointed to the dangers of 'test syndrome', in which testing troops become overly attached to their projects and want to see the best in them.[31] The results of this competition have hardly been borne out by later experience. In all conflicts involving the British Army since 1986 troops have pressed any available FN MAGs back into service at section level. In May 2003 the Ministry of Defence bowed to operational reality and announced that the FN Minimi – a 5.56mm weapon, but a proper light machine gun (squad automatic weapon – SAW in current jargon) – was to be issued throughout the Army.[32]

There was one further casualty of the SA80/LSW saga. The only new piece of weaponry needed by the infantry that was specifically mentioned in *Lessons* was a grenade launcher firing an area-effect grenade. At the time the only such weapon in general service was the US Army's M203 40mm under-barrel launcher, operated by a hi-lo pressure system. The British Army already used the M203 as an attachment to the SAS's M16 assault rifles. The M16 was, however, a conventional design with the magazine ahead of the grip. The M203 could not be easily adapted to the SA80, so the concept was allowed to wither.[33] As a result the Army had to rush rifle-launched grenades into service for the first Gulf War in order, 'to provide infantry platoons with a close assault weapon' – precisely the role that had been defined after the Falklands.[34]

There was, however, one less heralded but happier outcome to the Falklands small arms experience. In part, the Army's enthusiasm for the SA80-equipped marksman grew out of its long-standing suspicion of specialist snipers. 'The problem with using snipers,' one infantry officer observed, 'lies both in the difficulty of organising their training and also their deployment to the right spot'.[35] British snipers in the Falklands were issued with the L42 rifle, a re-bored 7.62mm conversion of the venerable Lee-Enfield design. These elderly weapons, further handicapped by sights prone to fogging, were shown to have reached the end of their useful life in the Falklands. Argentine snipers, albeit also equipped with vintage weapons such as the Argentine-made Mauser K98k or the American National Match M14, proved to be one of the enemy's few tactical assets. A trial for a replacement for the Lee-Enfield was held in 1982. The winner was a small English firm, Accuracy International, founded by a group of target shooters. The Accuracy International PM was adopted into service and has proved to be one of the best military sniper rifles in the world. Although the British Army did not succumb to the 'cult of the sniper', post-Falklands, snipers, at least in specialist units such as the paratroopers, came to be regarded as permanent and legitimate elements of the tactical mix. Army public relations even saw them as a potential 'hearts and minds' tool, as evidenced by the well known photograph of the 1990s showing a para-trooper sniper having his Accuracy International rifle aped by a stick-armed Kosovar child.[36]

Joint warfare

In many respects the command and control of operations in the South Atlantic was exemplary. Clear and decisive political leadership was provided by Margaret Thatcher as prime minister. Her small War Cabinet functioned smoothly. The diplomatic and military aspects of the campaign were prop-erly integrated. There was a strong partnership between the prime minister and the Chief of the Defence Staff (CDS), Admiral Sir Terence Lewin. Lewin presided over the deliberations of the Chiefs of Staff Committee, which met once or twice per day during the war, in a model fashion. His status had been increased only a few months before the war, making him a superior officer to the other chiefs rather than *primus inter pares*.[37] As a result military advice was channelled in an accurate and timely fashion to the political leaders; whilst the military leaders had a good grasp of the political considerations that were shaping operations. Both political and military leaders received enough good intelligence on which to base their decisions. Even in the most difficult circumstances this synergy between politics, war-fighting and intelligence passed the test of conflict. The most decisive act of the war, the sinking of the Argentine cruiser *General Belgrano* by the SSN *HMS Conqueror* on 2 May, demonstrated how well difficult politico-military decisions could be made in time-sensitive circumstances. British decision-making appears even more impressive when compared to the poor performance of the enemy.[38]

Unfortunately, the aspects of the system of command and control that worked least well were those that affected the land operations the most. The operational headquarters adopted for the war was the Joint Maritime HQ at Northwood in northwest London. Sir John Fieldhouse, overall commander of the operation, already operated from Northwood as C-in-C Fleet, as did the AOC 18 Group (Maritime) for the RAF. It was the army component that was added to the headquarters specifically for the Falklands.[39] Michael Rose, commander of 22 SAS in the Falklands, was scathing about the lack of influence this arrangement afforded to the tactical commanders during the land operation. 'I had seen for myself, he later wrote, 'how British lives had been unnecessarily lost in the Falklands because of political interference.'[40]

Britain was ill prepared for joint operations. This lack of preparedness had a number of long-term causes. The experience that Britain had garnered during the 1960s from operations such as the confrontation with Indonesia and the withdrawal from Aden in 1968 had largely atrophied. Since 1968 Britain had been committed to a European military role. In 1974 the incoming Labour government had insisted that residual 'out of area' capabilities should no longer be hidden under the NATO umbrella. The Army strategic reserve was broken up, amphibious forces were reduced and the capability for air dropping two parachute battalions was scrapped. This political pressure had reinforced the tendency of the services to concentrate on their own primary missions. It had also created divisions within the services. There was, for instance, a considerable distance between those parts of the Navy committed to the east Atlantic role under the Supreme Allied Commander Atlantic (SACLANT) and those committed to amphibious operations in northern Europe in the Supreme Allied Commander Europe (SACEUR's) area. These inadequacies and divisions were cruelly exposed by the Falklands.

One of the most glaring lessons of the conflict was the need to address the question of joint warfare. Yet the important strategic and budgetary calculations that had militated against this before 1982 were still as strong after the war. Both before and after the war Britain maintained the UK-Netherlands Amphibious Force in order to reinforce NATO's Northern Region. Post-1982 it was openly admitted that 'it has a value outside NATO, as has been seen by the Falklands conflict'. Yet this did not mean that the UK needed an 'all-singing, all-dancing amphibious capability'.[41] Joint warfare was thus a lesson both learned and ignored.

At the start of the war it soon became apparent that the readily available land striking force available for the operations would be inadequate. There were two formations with claims for inclusion in the operation, 3 Commando Brigade, a Marine unit specialising in amphibious warfare, and 5 Infantry Brigade, an Army unit that formed the UK's remaining 'strategic reserve'. 3 Commando Brigade would be the spearhead formation but it was too weak for the task of seizing back the islands by an opposed landing. In order to provide it with the necessary reinforcements, 5 Infantry Brigade was hollowed out by the transfer of 2 and 3 Para to the Marines. Subsequently it was decided that 5 Infantry Brigade would be needed as well.[42] This meant that the major Army formation committed to the war was a scratch organisation. It had no logistical organisation worthy of the name.[43] Its HQ group too was in poor shape. Under-manned and under-financed, it existed only in embryo form and had never conducted a full brigade exercise.[44] Its one remaining original unit, the 7th Gurkha Rifles, had to be reinforced by battalions of the Welsh and Scots Guards taken off ceremonial duties in London.[45] Brian Pennicott, the Royal Artillery commander in the conflict, noted that even as the forces sailed south there was an unfounded 'unspoken assumption' that 3 Commando Brigade would win the war and 5 Infantry Brigade would garrison the reconquered islands.[46]

The means by which this divisional expeditionary force made up of a heterogeneous collection of Marine commandos and Army battalions with their supporting units, was to be controlled, was not immediately clear. 3 Commando Brigade was a combat-ready formation. It proved relatively easy to 'blister on' the parachute battalions. The role and organisation of higher formations was more nebulous. Jeremy Moore was appointed to command the land operations. Moore was Major-General, Commando Forces. This was an administrative rather than a combat command, however. Although Moore was personally well qualified for the role he took on, his appointment was not the result of any pre-existing system.[47] In 2003, Sir Jeremy still had to give a long-winded explanation about his role in the chain of command.

> I had three jobs. The first one was what was called Major-General Commando Forces. I had to mount 3 Commando Brigade in the amphibious shipping to go to war. Second, I moved to Northwood after one week where I was Land Deputy to the Commander-in-Chief Fleet, who was also the commander of the taskforce. Thirdly, I went south and took over as commander Land Forces Falkland Islands from Julian

Thompson. … [In] sheer book terms I was senior to [Sandy Woodward], but that was irrelevant. We were co-equal task group commanders, though had it arisen (but it didn't) he would have been *primus inter pares*.[48]

In order to transport 5 Infantry Brigade to the South Atlantic the *QE2*, a large ocean-going liner, was requisitioned. The *QE2* was Moore's headquarters during the period between his leaving Northwood and arriving in the Falklands, the very time when the landing and the breakout was occurring. He had with him Brigadier Tony Wilson, the CO of 5 Infantry Brigade, but his relations with Brigadier Thompson who was fighting the battle were patchy. The satellite communications system Skynet 2B proved thoroughly unreliable – its replacement was accelerated in the aftermath of the war.[49] As a result, Thompson and Moore rarely talked to each other, yet from 20 May onwards signals came down the chain of command to Moore rather than to Thompson.[50]

In retrospect the Army concluded that this aspect of the operation was poor. 'Delegating the planning of operations to someone who eventually has to execute those plans always has considerable merit', wrote Edwin Bramall, the CGS during the war,

> but, in practice and as events progressed, it became increasingly difficult to reconcile these two roles; and there turned out to be about 72 crucial hours, after the initial bridgehead had been established, when the Land Force Commander's presence was badly needed ashore at San Carlos and yet Moore was still on his way down by sea. … He was, therefore, out of touch with the battle and unable to provide the immediate impetus and direction which was required.

'With hindsight', Bramall continued,

> Moore should have handed over to the Army (or another Royal Marine) adviser somewhat earlier, or somehow travelled more quickly and thus would have been able to advise Woodward during the final deployment and approach to the landing area and, much more importantly, to command ashore at an earlier stage.[51]

Thompson's conduct of the initial battle was, and remains, controversial. He himself has engaged in some honest self-criticism about his tactics, particularly with regard to the battle of Goose Green. A general feeling in the Army that they, rather than the Marines, should have led the operation, was balanced, however, by a recognition that 3 Commando was the only unit that constituted a credible assault force in 1982. It was this assault element that came to the fore in the post-Falklands reforms. By the end of 1982 the postwar review had concluded that 5 Infantry Brigade should be reconstituted as a much more battle-ready formation. An artillery regiment, an Army Air Corps squadron and logistic support units were added to its order of battle. RAF C-130 Hercules transport aircraft were upgraded so that the Brigade would be capable of carrying out a two-battalion airborne assault if necessary. In any future conflict the Army was to have a capability that would rival that of the Marines, who would no longer be the automatic choice as the spearhead formation.[52] 5 Airborne Brigade, as the enhanced formation became known, became fully operational at the beginning of 1986.[53] Laying claim to primacy in the assault role thereafter, the Army even made an attempt to seize control of the Marines from the Royal Navy, albeit with little success.[54] They declined, however, to become more like the Marines. The Army rejected the Marines brigade-based logistics model, which had proved so successful in the Falklands, on the grounds that it would burden BAOR commanders 'with excessive responsibility for logistic support'.[55]

It was a relatively straightforward decision for the Army to upgrade one of its formations so that it would not be found wanting in future in comparison to a naval force. The issue of the future command and control of joint operations evoked a much more ambivalent response. In March 1983 Admiral Woodward publicly called for the creation of a permanent joint force headquarters, to address the weaknesses that been apparent in the Falklands. Although it was hard to argue against the evidence of the campaign that such an organisation was in some ways desirable, the Army treated the proposal with caution.[56]

A tri-service Permanent Planning Group (PPG), headed by an Army colonel, was created. This planning cell was attached to HQ South East District and came formally under the District GOC. The PPG was supposed to form the nucleus of a dormant Major-General's command that could be expanded into a JFHQ if necessary. In the meantime its primary role was to give more thought to the JFHQ concept. The cell did not get very far. By the late 1980s it had evolved into the Joint Force Operations Staff (JFOS), with an establishment of thirteen officers and four NCOs, still with a colonel as chief of staff. The GOC SE District was 'double-hatted' as Commander, JFOS. In 1990 the serving GOC SE District, Sir Peter de la Billière, was appointed to lead British forces in the first Gulf War, the first major expeditionary operation since the Falklands. Yet this appointment was more a result of political manoeuvring within the armed services for a plum command than any preconceived notion that Commander, JFOS would be the automatic choice for the job.[57] Indeed, de la Billière later made clear that he had had little to do with the JFOS staff. He did not regard it as the nucleus of his command and he had taken little interest in its activities.[58]

De la Billière himself considered that the most important thing about the Falklands experience was that he himself had 'got on well' with the Royal Navy whilst garrison commander in the mid-1980s, making them less likely to veto his appointment in 1990. In his view it was the experience of the post-war garrison that had done most for inter-service cooperation rather than any lessons that might have been drawn from the war itself. 'My time in the Falkland Islands', de la Billière wrote,

> had taught me that tri-service command is a peculiar art, difficult to manage until one is used to it. The Army, Navy and Air Force all have their own procedures and ways of doing things; it therefore needs a combination of tact and firmness to make them work together in harmony. Rather than issue sweeping orders and directives, one has to bring all three services into line by gentle but firm manipulation.

General de la Billière believed that he had succeeded as a 'purple' commander on the basis of his own personal abilities and experiences rather because of any Army system for joint operations.[59]

A similar sense of desultoriness with regard to system was felt elsewhere in the Army. In the immediate post-Falklands glow the Directorate of Military Operations had committed itself to a 'pattern of amphibious exercises which will involve Army formations', in particular 5 Brigade.[60] One major joint force exercise, *Saif Sareea*, was held in Oman in 1986. The next such exercise in the programme, *Purple Victory*, was to have had the fully reconfigured 5 Airborne Brigade at its heart. The plans were, however, drastically reduced on grounds of cost, even before the exercise was overtaken by events in the Gulf.[61]

On the eve of the Falklands War, plans were being made to decrease rather than increase joint force activity. In 1981 the decision had been taken to close the Joint Warfare Wing (JWW) of the National Defence College. This was a victory for the single services who did not wish to waste officers or money on such peripheral activities. Instead, 'any such training thought to be essential would be carried out on a single service basis'. This decision looked less than far-sighted in the spring of 1982, and the CDS reprieved the JWW and ordered a further review. As a result the Joint Warfare Staff was created in 1983 at the HQ, UK Land Forces. It was relocated alongside the Marines at Poole in 1985. The main role of the new staff was running courses on combined operations. Yet there was no evidence that attendance at these courses was necessary for a successful career in the Army. Members of the JWS had to become 'accustomed to the occasional puzzled question, "Aren't you the people at Old Sarum?" ' – the Joint Warfare Establishment had been based at Old Sarum between 1963 and 1979, when it was closed. The JWS's own commander acknowledged that 'there are those who fear this is as yet another turn in the tide of the UK approach to joint warfare', but argued that 'the lessons learned in the South Atlantic were powerful ones'.[62] The fact remained that what had once been a lieutenant-general's posting was now held by a colonel.[63]

There is a striking contrast between the very cautious reforms enacted after the Falklands War and the burgeoning of 'jointery' in the 1990s, leading to the creation of Permanent JFHQ at Northwood and the Joint Rapid Reaction Forces. Although it can certainly be argued that some of the groundwork for these organisations was laid in the 1980s, they were, in fact, a response to the strategic environment that was created by Anglo-American operations in the Gulf and the end of the Cold War. It was hardly accidental either that the Army was converted to 'expeditionary warfare' when the 'heavy' Army became the core of the expeditionary force as was the case in the Gulf.

The test of battle

There is little doubt that the Army benefited from the relatively short time that elapsed between the Falklands and its next expeditionary war in the Gulf, just eight years later. Experience of real operations had not wasted away as it had in the twenty-six year gap between Suez and the Falklands. Although the type of 'teeth units' – two armoured brigades – deployed to the Gulf was quite different from those found in the Falklands, some of the 'backroom' problems such as logistics were quite similar. 'We have been struck', noted a House of Commons Defence Select Committee enquiry, 'by the number of times we have been told of lessons learned and improvements or adjustments made after Operation *Corporate* [i.e. the Falklands] which bore fruit in the Gulf'.[64] The general conclusion reached after the Falklands that training had to be as realistic as possible had not had time to fully dissipate.[65] The war's impact was felt more in practical and technical matters than in any grand reform.

In the longer perspective the Falklands War can be seen not so much as an event that fundamentally changed perceptions within the Army but as part of a much wider process. At the beginning of the 1980s, the Army seemed to be faced by the polarities of deterrence and low-intensity operations.[66] Both the changes occurring to the British stance in northwest Europe during the 1980s and the lessons of the Falklands War offered an alternative way forward; a belief that at the heart of the Army's mission lay the search for victory in conventional conflict. The intellectual ferment of the 1980s taught this through doctrinal debate; the Falklands reinforced the same conclusion by hard and bloody experience. The conclusion reached by Hew Pike, commander of 3 Para in the Falklands, an officer who finished his career as a full general, is particularly revealing. 'What is perhaps most interesting', Pike wrote a decade after the end of the war,

> are the unchanging lessons of such battles, all of which would have been familiar to our fathers & grandfathers. Yet no-one, in either brigade, from brigadier downwards, had any experience of such actions, and no amount of history is a substitute for experience.[67]

Thompson himself agreed with this analysis. He commented that, 'we learned – or rather re-learned – a huge number of lessons (you never learn new lessons, I think, you just re-learn old ones)'.[68] The collapse of the Soviet Union and the first Gulf War changed the expected theatre of operations, but reinforced this process of re-learning.

As the Army moved into the expeditionary age of the 1990s, many of the lessons, particularly regarding joint warfare, that had been flagged in the aftermath of the Falklands suddenly seemed more relevant. By this stage, however, the Army had seen action in a number of other conflicts. A story told by Michael Rose, like Pike another Falklands commander who reached the rank of full general, perhaps best encapsulates the longevity of the Falklands experience. As commander of the UN in Bosnia during the Yugoslav civil war he had occasion to visit a brigade of the Serbian army. To his surprise the unit's medical officer was well informed about the Falklands. He and Rose had fallen into a conversation about its lessons when, 'the arrival of the brigade commander cut our discussion short, and we started to talk about the war in Bosnia'.[69] Although the Falklands War lived long in the memory, memory always had to play second fiddle to pressing current concerns.

Notes

1 Cmnd. 8758, *The Falklands Campaign: The Lessons*, December 1982.
2 Field Marshal Lord Bramall, 'Task Force Falklands', in Linda Washington (ed.) *Ten Years On: The British Army in the Falklands War,* London: National Army Museum, 1992, pp. 5–18.
3 Cmnd. 8758, December 1982.
4 Lawrence Freedman, 'British Defence Policy after the Falklands', in John Baylis (ed.) *Alternative Approaches to British Defence Policy* London: Macmillan, 1983, pp. 62–75; Andrew Dorman, Michael Kandiah and Gillian Staerck (eds) *The Nott Review*, London: Institute of Contemporary British History, 2002.
5 Bruce Watson and Peter Dunn (eds) *Military Lessons of the Falklands War: Views from the United States* (Boulder CO: Westview, 1984).
6 Julian Thompson, *No Picnic* (London: Leo Cooper, 3rd edn, 2001); *idem.*, 'Falklands: With Hindsight', *Army Quarterly*, 122 (1992), pp. 263–6; *idem.*, 'The Falklands, 1982: War in the South Atlantic', in Julian Thompson (ed.) *The Imperial War Museum Book of Modern Warfare: British and Commonwealth Forces at War, 1945–2000* (London: Sidgwick & Jackson, 2002), pp. 289–308.
7 Sandy Woodward, *One Hundred Days: The Memoirs of the Falklands Battle Group Commander*, London: HarperCollins, 2nd edn, 2003.
8 Woodward, *One Hundred Days*, pp. 124–6.
9 Thompson, *No Picnic*, pp. 17–18.
10 Colin McInnes, *Hot War, Cold War: The British Army's Way in Warfare* (London: Brassey's, 1996), pp. 60–8; John Kiszely, *The British Army and Approaches to Warfare since 1945* (Strategic and Combat Studies Institute, Occasional Paper No. 26, 1997), pp. 26–8. Kiszely won a Military Cross as a Scots Guards officer in the Falklands.
11 Editorial, *British Army Review*, 72 (December 1982).
12 Evidence before the House of Commons Defence Select Committee on 8 May 1991. *House of Commons, Session 1990–91*, 287-i, 17 July 1991.
13 Freedman, 'British Defence Policy after the Falklands', pp. 62–75.
14 Alex Danchev (ed.), *The Franks Report*, London: Pimlico, 1992.
15 Cmnd. 8758, December 1982.

16 Cmnd. 8758, December 1982.
17 Evidence of Miss Margaret Aldred (Director of Defence Policy, MoD) to the Defence Committee, 20 October 1993, *House of Commons, Session 1993–94*, 43, 25 May 1994.
18 *House of Commons, Session 1986–87*, 345-i. A unit-by-unit assessment of the war, culled from regimental and unit magazines can be found in Derek Oakley, *The Falklands Military Machine*, Speldhurst: Spellmount 1989, pp. 54–79.
19 Thompson, 'Falklands: With Hindsight', pp. 263–6.
20 Edward Fursdon, *Falklands Aftermath: Picking up the Pieces*, London: Cooper, 1988.
21 Thompson, *No Picnic*, p. 143.
22 *House of Commons, Session 1986–87*, 345-i.
23 W.J. Tustin, 'The Logistics of the Falklands War, Part 2', *Army Quarterly*, 114 (1984), pp. 398–411.
24 Cmnd. 8758, December 1982.
25 Information provided by the Ministry of Defence to the House of Commons Defence Select Committee, *House of Commons, Session 1992–93*, 728, 9 June 1993.
26 G.J. Phillips, 'The Influence of the SA80 on Low Level Tactics', *British Army Review*, 75 (1983).
27 Lecture by S.J. Oxlade (MoD Infantry Trials and Development Unit), 'The Importance of the Battlefield Mission in the Small Arms Trials Process – A Personal View' (2000).
28 D.F. Allsop and M.A. Toomey, *Small Arms: General Design*, London: Brassey's, 1999, pp. 64–5.
29 F. Myatt, 'The Light Machine Gun in the British Army', *British Army Review*, 70 (April 1982), pp. 56–9.
30 Oxlade, 'Battlefield Mission'.
31 *House of Commons, Session 1992–93*, 728, 9 June 1993.
32 The Minimi had been purchased for use by the SAS in 1988. Max Hastings, 'Our Soldiers Pay the Price for Bureaucratic Folly in Whitehall', *Guardian*, 31 July 2004.
33 *House of Commons, Session 1986–87*, 228, October 1987. More recent versions of the M203 are compatible with bullpups.
34 Memorandum submitted by the Ministry of Defence to House of Commons Defence Committee, June 1991, *House of Commons, Session 1990–91*, 287-i, 17 July 1991.
35 Phillips, 'Low Level Tactics'.
36 Martin Pegler, *The Military Sniper since 1914*, Oxford: Osprey, 2001, pp. 49–51.
37 John Nott, *Here Today, Gone Tomorrow: Recollections of an Errant Politician*, London: Politico's, 2002, p. 247.
38 *The Falklands Witness Seminar*, Joint Services Command and Staff College, 5 June 2002 (Strategic and Combat Studies Institute, Occasional Paper No. 46, 2002).
39 Bramall, 'Task Force Falklands', pp. 5–18.
40 Michael Rose, *Fighting for Peace: Lessons from Bosnia*, London: Warner, 1999, p. 77; *idem.*, 'Advance Force Operations: The SAS', in Washington, *Ten Years On*, pp. 55–60.
41 Evidence given to House of Commons Defence Select Committee by Sir Edwin Bramall, *House of Commons, Session 1984–85*, 37, 23 May 1985.
42 *The Falklands Witness Seminar*, pp. 40–1, 50–1.
43 Tustin, 'The Logistics of the Falklands War, pp. 398–411.
44 Bramall, 'Task Force', pp. 5–18.
45 Nott, *Here Today*, pp. 305–6.

46 Brian Pennicott, 'The Gunners', in Washington, *Ten Years On*, pp. 49–54.

47 *House of Commons, Session 1986–87*, 345-i.

48 *The Falklands Witness Seminar*, p. 6.

49 *House of Commons, Session 1986–87*, 345-i.

50 Julian Thompson, 'Falklands: With Hindsight', pp. 263–6.

51 Bramall, 'Task Force Falklands', pp. 5–18. Bramall was Chief of the General Staff between July 1979 and October 1982.

52 Cmnd. 8758, December 1982.

53 *House of Commons, Session 1986–87*, 345-i.

54 *House of Commons, Session 1993–4*, 655, 29 September 1994.

55 Minutes of House of Commons Defence Select Committee, October 1987, *House of Commons, Session 1986–87*, 345-ii.

56 Evidence given by C.T. McDonnell (AUS General Staff, MoD) to House of Commons Defence Select Committee, 22 February 1984, *House of Commons, Session 1986–87*, 345-ii.

57 Peter de la Billière, *Storm Command: A Personal Account of the Gulf War*, London: HarperCollins, 1992, pp. 11–13.

58 Evidence before the House of Commons Defence Select Committee on 8 May 1991. *House of Commons, Session 1990–91*, 287-i, 17 July 1991.

59 De la Billière, *Storm Command*, pp. 16, 57.

60 Evidence given by Colonel H.A. Woolnough to House of Commons Defence Select Committee, 22 February 1984, *House of Commons, Session 1986–87*, 345-ii.

61 *House of Commons, Session 1990–91*, 287-i, 17 July 1991.

62 Colonel A.R. Jones, 'The Joint Warfare Staff', *British Army Review*, 87 (December 1987), pp. 41–4.

63 *House of Commons, Session 1986–87*, 345-i.

64 *House of Commons, Session 1990–91*, 287-i, 17 July 1991.

65 Jonathan Bailey, 'Training for War: The Falklands, 1982', *British Army Review*, 73 (1983), pp. 21–30; Hew Pike, 'The Army's Infantry and Armoured Reconnaissance Forces', in Washington, *Ten Years On*, pp. 40–8.

66 Sir Edwin Bramall, 'British Land Forces: The Future', Lecture to RUSI, 17 February 1982. Published in *RUSI Journal*, June 1982.

67 Pike, 'The Army's Infantry and Armoured Reconnaissance Forces', pp. 40–8.

68 *The Falklands Witness Seminar*, p. 47.

69 Rose, *Fighting for Peace*, p. 207.

7   The Gulf War, 1990–1

Colin McInnes

One of the key questions underpinning this collection of essays is whether colonial campaigns and counter-insurgency operations aided the Army's ability to fight major wars, or undermined it.[1] This essay addresses the 1990–1 Gulf War and specifically whether the two decades prior to 1990–1 – a period of relative inactivity regarding major war, but one where the small war[2] of the Falklands and counter-terrorist operations in Northern Ireland loomed large – prepared the Army for its first use of an armoured division in war since 1945. It is, however, worth noting that behind this question is an assumption about the nature of war which held sway during this period, and indeed throughout the period after the Second World War. That assumption is that 'small wars', and in particular counter-insurgency and counter-terrorist operations, require different skills and present different challenges to armed forces than the problem of major war. For the post-war Army, the conduct, strategy and tactics of these wars was fundamentally different, and a failure to recognise this flirted with disaster. Even the nomenclature was different, few if any being dignified with the term 'war' – Malaya, the Falklands and Northern Ireland being amongst the more obvious examples. This assumption is not always shared by other armed forces, but is a *leitmotif* of British military thinking in the post-war era.

The orthodox interpretation of British military history in the Cold War period is that the Army has been relatively successful in small wars.[3] When this interpretation is coupled to the above assumption that small wars require different skills and approaches from major wars, the question of whether this excellence in small wars may be paid for at a cost in terms of the Army's ability to conduct major operations is perhaps an obvious one to ask. But this question acquired a new dimension in the aftermath of the combined events of the end of the Cold War and the 1990–1 Gulf War. During 1990–1 the Options for Change 'exercise'[4] began the process of reducing the size of Britain's military for the post-Cold War world. But this process did not finish with Options for Change; rather it continued throughout the decade, arguably culminating in Labour's 1998 Strategic Defence Review.[5] The Army of the future would clearly be smaller than that of the Cold War, but conceivably used on a more regular basis – as events in the Gulf and former Yugoslavia quickly demonstrated. The question faced by the Army during this period was whether to exploit its long-standing reputation for excellence in small wars and prioritise the emerging role of peace support operations, or to remain a 'player' amongst NATO military circles by retaining a capability for fighting a major war by using technologically adept armoured forces. The Army chose to have its cake and eat it. By the mid-1990s it was arguing that, in retaining a capability for high intensity warfare, it would be able to meet its most challenging task of fighting a major war; and that, because it could meet this task, it could also manage less demanding tasks such as peace support operations. The assumption here was that, although the two tasks were different, they were not sufficiently different as to invalidate each other in an era of constrained resources. The small post-Cold War British Army could excel at both peace support operations and major wars and the evidence for this lay in the 1990–1 Gulf War: despite not having fought a major war since Suez, despite fighting a series of 'small wars' ranging from counter-terrorism in Northern Ireland through counter-insurgency in the former empire to the minor conflict of the Falklands, the focus on BAOR meant that the Army was nevertheless able to deploy and fight an armoured division with apparently startling success in the Gulf War, whilst still being able to engage in small wars.

Thus the 1990–1 Gulf War became a crucial example of two contestable phenomena. The first of these was that in the post-Cold War world large-scale conventional conflicts could occur and that the British Army might be involved in their prosecution, thereby enabling the UK to 'punch above its weight' diplomatically. The second of these contestable phenomena was that the ability to undertake such tasks need not be undermined by more frequent small-scale, low intensity operations. These were political arguments which were used in the struggles over the size and shape of the post-Cold War Army, but at their heart lay an answer to the question underpinning this chapter.

The shadow of the past: the Army in the 1970s and 1980s

Assessing whether the 'small wars' of the 1970s and 1980s assisted or undermined the Army's ability to conduct a major war such as the 1990–1 Gulf War is somewhat difficult because, despite the occurrence of 'small wars' (including both the Falklands and Northern Ireland), the focus for much of this period was on the prospect of a major war with the Soviet Union. From Denis Healey's reviews of the late 1960s onwards, the thrust of British defence policy had been to prioritise NATO commitments. This was reinforced by the Mason Review of 1974–5 and the 1981 Nott Review.[6] For the British Army, this meant a focus upon BAOR and the defence of NATO's Central Front. This was where the bulk of the Army's best equipment was deployed, this was where almost all of the heavy armoured units were stationed, and this represented by far the greatest concentration of British forces outside Great Britain *including* Northern Ireland. Of course, from time to time other conflicts did receive more attention. There were certainly periods during the 1970s when the Army's focus appeared to be more on Northern Ireland than BAOR (though with the introduction of police primacy in the late 1970s this level of attention was reduced), while the brief period of the Falklands Conflict understandably drew attention away from BAOR. Nevertheless, through the period as a whole the Army's focus, like that of British defence policy as a whole, was on its NATO commitments in general and the Central Front in particular. 'Small wars' were seen as distractions, some more temporary than others. Some were barely acknowledged – such as the Dhofar campaign.[7] As a result, the Army was probably better prepared intellectually and materially to fight a major war than it was a small war such as the Falklands. This is not to say that the small wars of the 1970s and 1980s did not have an impact on the Army – they did and some of this will be discussed below. Rather it is to suggest that the ability of these small wars to undermine the Army's preparedness for major war was constrained by the consistent focus on BAOR.

There is, however, another dimension to this focus on BAOR which might have undermined the Army's preparedness for major war, and that is the focus on a single scenario war. BAOR was not designed to fight a major war in general, but to fight a very specific war: one in northern Germany on ground extensively reconnoitred, using forces many of which were already in place and whose reinforcement had been carefully planned, on the defensive, against a particular enemy whose strategy, tactics and force structure had been extensively examined, and in the context of possible nuclear use. One is struck reading articles by Army officers through the 1970s and 1980s by the fact that the possibility of a major war *outside* this context seemed to them to be as unlikely as the end of the Cold War. Thus although the focus on BAOR assisted in preparing the Army intellectually and materially for a major war, the focus was a narrow one which did not seriously entertain the prospect of the Army sending a large armoured formation to fight a war overseas against an enemy other than the Warsaw Pact.

To enable useful analysis, this chapter adopts a six-part framework covering various aspects of the Army. This framework is not intended to be exhaustive, covering all aspects of the Army, but rather illustrative. Its purpose is to enable an analysis of the extent to which developments in the 1970s and 1980s, including both 'small wars' and preparations for major war on a central front, affected the Army's preparedness for the 1990–1 Gulf War and the manner in which it approached operations in the Gulf.

Human and material factors

One of the key themes running through the history of strategic thought is the relative balance of human and material factors in determining the outcome of battle and war. For some, such as Ardant du Picq, moral factors are key determinants; for others, such as J.F.C. Fuller, material factors including technology are the more important. The British Army, and NATO more generally, was not in a position to challenge Soviet quantitative superiority, though it did hold the potential for a technological lead. But to what extent should this be emphasised over moral factors such as leadership, initiative and professionalism?

BAOR was deployed in the most technologically sophisticated and machine-intensive theatre of war. Indeed for a peacetime deployment, NATO's armies on the Central Front demonstrated remarkable technological sensitivity. Despite this, however, the British Army remained wedded to an emphasis upon more human qualities. Unlike the majority of NATO armies, BAOR during this period was a fully professional army. Further, criticisms of its equipment – of which there were many, particularly in the 1970s – were offset by praise for the quality of its troops. Northern Ireland provided experience of 'real soldiering' as opposed to peacetime training, but the qualities required there were of individual initiative, morale, courage and leadership. A 'platoon commander's war', Northern Ireland offered little scope for technology but allowed the Army to develop and exploit more human qualities. Nor was the land element of the Falklands Conflict equipment intensive. Rather what emerged was how critical a high *esprit de corps* could be.

Training

The picture which emerges concerning the extent and effectiveness of the Army's training during the 1970s and 1980s is unclear. To the extent that BAOR was a professional standing army with dedicated training facilities (including BATUS, the battalion-level training facility in Canada) and an established series of exercises, the Army appeared well trained and prepared for high intensity armoured warfare. Further, Northern Ireland provided a degree of combat experience for generations of infantrymen and other combat and support arms, while the success in the Falklands could be interpreted as evidence of a successful training regime. However, BAOR's training budget was regularly at risk to budget cuts during this period, most notably in the late 1970s when it contributed to a crisis in morale; the training and experience provided by Northern Ireland was limited in both the type of operations involved and the type of troops;[8] and there is some concern that the evident high levels of professionalism in the Falklands were in no small part due to the high proportion of elite forces used, notably paratroopers and Royal Marine commandos.

Value and utility

Perceptions of the value and utility of the armed forces may vary not only over time but across sections of society. It is also important to distinguish between perceptions of value – that is, that the armed forces are particularly valued at one moment in time to the UK, perhaps because of a heightened threat level – and utility, by which I mean that the British armed forces are seen as being particularly good at their job and can be used with confidence. The 1970s saw the value of the armed forces reduce somewhat with the withdrawal from Empire almost complete and detente reducing Cold War tensions. Northern Ireland offered a reminder of the Army's value, though events such as Bloody Sunday also raised questions over its utility. The early 1980s, however, saw perceptions of both the value and utility of the armed forces in the UK begin to rise, both because of the heightened tension of the second Cold War, but also in the wake of the successful Falklands conflict. The Thatcher government also placed the armed forces at the centre of its attempt to rehabilitate Britain's world standing, not least with the purchase of the state of the art Trident strategic nuclear deterrent, but also more generally with increased defence expenditure[9] and a higher political profile for defence. With reduced tensions in the late 1980s and then the end of the Cold War, however, the value of armed forces, at least in the East-West context, looked less certain.[10]

Special forces

One of the features of the 1980s for the Army was the increased attention paid to special forces. Although such forces had been an almost constant feature of the Army's order of battle since 1945,[11] comparatively little attention had been paid to their role both publicly and to a certain extent within the Army itself. This began to change with the 1980 siege of the Iranian embassy in London. Its denouement of an SAS assault was not only hugely successful politically and militarily, but was dramatically shown live on television. Public interest in the SAS was immediately kindled, to be further inflamed by its role in the Falklands War. In both instances, public fascination was allied to political realisation that special forces might prove a useful tool. The role of such forces in Northern Ireland also increased from the 1970s, with the SAS appearing to take on the role as the lead offensive arm of the security forces. This role proved somewhat more controversial, however,[12] perhaps the most problematic incident being the killing of members of an IRA active service unit in Gibraltar. As Hew Strachan commented on the incident, 'What alarmed many was the assumption of powers of execution by armed servicemen in civilian clothes whose policy was to shoot on sight without warning'.[13] Nevertheless, by the end of the decade special forces, and particularly the SAS, had become a highly visible and important element of the Army. But the role of the Special Forces appeared most pertinent to small wars such as Northern Ireland and the Falklands, while its impact in major combat operations was less obviously central. Unlike Northern Ireland or the Falklands, the SAS did not figure prominently in BAOR's thinking about how to fight a major war, the attention instead being focused upon more traditional and significantly larger armoured units.[14]

Contingency planning

Although contingency planning was undertaken throughout these two decades for a variety of possible crises, the overwhelming focus of planning was on the possibility of a war against the Warsaw Pact. This included the allocation of units to specific tasks and sectors, detailed logistical work, thinking about the deployment of reserves, developing operational procedures for working with allies and thorough reconnaissance of the likely ground on which BAOR would fight. In comparison, work on the possibility of a conflict over the Falklands was scant with no adequate plans and a force which, to critical eyes, was cobbled together. Nor had there been detailed planning for a deployment to Northern Ireland in 1969, nor indeed to Dhofar. But to a certain extent this misses the point. The British Army's traditional role of imperial policeman had meant that it had relied less on detailed planning than on an ability to improvise. In a more charitable light, its experience suggested that detailed plans drawn up in advance rarely matched the reality of a crisis on the ground, and it was therefore better to have a capability which could be readily and flexibly deployed than a pre-ordained series of instructions. In this sense BAOR and the 1970s and 1980s were perhaps unusual in that the British Army did prioritise detailed planning based upon the contingency of a Warsaw Pact attack, and that when other crises did emerge requiring a military response, the ad hoc nature of this response was very much in tune with British military traditions.

Doctrine

As Brian Holden Reid has pointed out, the British Army has no tradition of thinking about war or of relying upon a military doctrine.[15] Although there is evidence of some written military doctrine prior to this period, much of this was tactical rather than operational or strategic in nature. What operational or strategic doctrine it did possess tended to be informal (a 'community of knowledge') rather than formally constituted.[16] And although there is ample evidence of individual officers engaging in and writing on doctrinal matters and on strategic theory (including perhaps most notably Fuller and Liddell Hart), there is no sense of an organisation which encouraged such thinking, or of an *institutional* structure wherein such thinking took place. As a result the Army had been (sometimes fairly, sometimes unfairly) portrayed as more interested in 'huntin', shootin' and fishin' ' than in reading and intellectual engagement with abstract theory over their chosen profession. As Lord Tedder infamously quipped, the British Army suffered from 'an excess of bravery and a shortage of brains'.[17]

During the 1980s however this began to change quite fundamentally. Thinking about doctrine became not only much more central to the Army but increasingly institutionalised with its own structures. What began in the early 1980s as a 'ginger group' under the auspices of BAOR commander General (later Field Marshal) Sir Nigel Bagnall, evolved by the end of the decade into a formal group charged with the development of Army doctrine, a major operational level doctrinal publication intended to inform all Army operations,[18] and the introduction in 1988 of a high-level staff course (the Higher Command and Staff College, HCSC) co-located with the Army Staff College at Camberley.[19]

Together with this institutional change, the direction of doctrine changed. Prior to the 1980s, the British Army's approach to operations – particularly on NATO's Central Front where BAOR was stationed – had been one of a series of tactical level engagements exploiting prepared ground, firepower and the natural advantages of the defensive. Although forces were mobile, they would use this mobility at the tactical level to move from one prepared position to another within a series of killing zones. This operationally static, firepower-oriented approach was consistent with that adopted by Montgomery in the Second World War and had informed BAOR's approach to operations throughout the Cold War period. Under Bagnall and his successor, General Sir Martin Farndale, BAOR and subsequently the Army as a whole adopted a more manoeuvrist outlook. This new approach, under-pinned by doctrine, emphasised: fighting large units, including 1(BR) Corps, as a whole at the operational level, rather than as a variety of smaller units in a succession of smaller battles; seizing the initiative rather than reacting to the enemy's moves; emphasising the movement of mass against key points, rather than masses of movement in a series of relatively uncoordi-nated battles; mission command whereby individual commanders were allowed greater freedom within an overall concept of operations, thereby speeding up the decision making cycle and allowing the initiative to be retained; and greater cooperation between land and air elements, viewing them as two dimensions of a single campaign.[20]

Operation Granby: the 1990–1 Gulf War

Having established a framework for comparative analysis in the section above, this section then proceeds to use this framework to examine British performance – and especially the performance of 1st Armoured Division, the main British Army fighting unit – in the 1990–1 Gulf War.[21]

Human and material factors

1st Armoured Division deployed the most modern equipment available to the British Army, including Challenger main battle tanks and Warrior infantry fighting vehicles. Although only two brigades strong, the division gave every appearance of being well equipped, deploying considerable fire-power and mobility. Nevertheless the performance of this equipment was somewhat patchy. A number of systems were identified as 'battle winners' giving the British division clear advantages. Amongst these were TOGS (Challenger's observation and gun sight) and multiple-launch rocket system

(MLRS), both relatively new systems. The performance of the satellite navigation system fitted to armoured vehicles in the Gulf was also widely praised. Its value stretched beyond simply allowing forces to navigate accurately across the featureless desert (in itself no mean advantage). Satellite navigation meant that movement of large formations at night was made much simpler, while tactical commanders were able to focus on fighting the battle rather than worrying where they (and other friendly forces) were. Most stunning was the use made of satellite navigation by the Staffords in attacking the Iraqi position codenamed PLATINUM: the ability to navigate with extreme accuracy meant that the Staffords could manoeuvre completely around the Iraqi position and attack it from an unexpected direction. Warrior infantry fighting vehicles performed well (some communication problems aside), though Challenger proved somewhat more equivocal. In its favour was its speed, protection and ability to engage targets at extreme range (up to 3,000 metres). More problematic were its ergonomics, systems integration and reliability. Although Challenger managed 95 per cent availability throughout the operation, a huge improvement on the figures achieved in Germany, this was only possible because of the very large number of spares held forward enabling prompt maintenance. Less satisfactory performances were achieved by the Lynx attack helicopter whose capabilities proved limited, and 16/5 Lancers' medium reconnaissance vehicle which proved to be vulnerable to enemy action, and whose petrol engines complicated the fuel resupply (all other armoured vehicles ran off diesel). Some of the artillery, the logistics resupply and the FV 432 vehicle found it difficult to keep up with the speed of the advance. Finally, existing logistic provision and transport capabilities were demonstrated to be extremely weak. The two brigades required almost a corps' worth of support, while adequate transport was lacking in both numbers and quality.

The overall assessment of equipment performance is therefore something of a mixed bag. The newer equipment tended to perform well (particularly that developed by the US such as MLRS and the satellite navigation system), while older equipment such as the FV 432 and the Midge drone were not satisfactory for this type of war. What is perhaps significant is that, even after the build-up and new equipment programmes of the 1980s, the British Army's equipment was, in general terms, only satisfactory, and that although it was better equipped than some of the coalition forces, it fell short of the standards set by the Americans.

In contrast to the somewhat equivocal performance of its equipment, the British Army considered that the individual qualities of its soldiers were extremely high and a key component in its success. The high morale and professional competence of personnel receive considerable attention in almost all official assessments of the Army's performance. For example, the 1992 Statement on the Defence Estimates noted that 'The commitment and professionalism of the armed forces and of the civilians who supported them were a key factor in the effectiveness of the British contribution in the Gulf.'[22] Similarly the House of Commons Defence Committee concluded that

> the principal lesson of Operation Granby is that the Armed Forces of the United Kingdom depend now, and will continue to depend, on the skill, dedication and courage of individual Servicemen and women. No amount of equipment, planning and organisation can replace that priceless asset.[23]

Therefore, despite the initial emphasis upon material factors, not least from the intense media coverage of the war which emphasised the new technologies being used, the overall result appears to have vindicated the British emphasis upon more human factors. This is not of course to say that material factors are unimportant to the British Army; rather it is to say that Operation Granby appeared to demonstrate to the British Army that in hot war as well as cold, in high intensity, high technology conflicts as well as low intensity, human factors remained vitally important.

Training

The first of the two brigades to arrive in the Gulf – 7 Armoured Brigade, or the Desert Rats – appeared far from ready to engage in combat operations, despite two of its three battle groups having just returned from a period of formation level training at the BATUS facility in Canada. Once in the Gulf however, their commander, Brigadier Patrick Cordingley, began an intense series of exercises, including live firing exercises to bring them up to peak levels. As the Secretary of State for Defence commented after the war

> The very remarkable military achievement ... owes a lot to the opportunity for intensive training and live firing practice that the troops and the armoured regiments were able to enjoy during the period before the conflict started.[24]

Training for other troops began in Germany, though this was limited to individual and battalion level. Major-General Rupert Smith, commander of 1st Armoured Division, commented that 'much training obviously had to continue after deployment to Saudi Arabia', a point reaffirmed by Brigadier Cordingley.[25] This suggests a general lack of combat readiness prior to deployment. Although this may to a certain extent be explained by the naturally lower levels of readiness experienced during peacetime, particularly with the end of the Cold War, the lack of live firing exercises and field exercises of large units were major deficiencies which had to be rectified. As the House of Commons Defence Committee commented

The relatively unconstrained training and exercising which the forces were able to undertake in late 1990 and early 1991 were luxuries rarely if ever enjoyed in peacetime conditions. A full armoured Division exercise, with unfettered live firing, is wholly impossible in European conditions.[26]

This intense period of training appeared to be extremely successful, not least in impressing American observers of the capability of the British forces.[27] That the British Army was able to bring its troops up to such a high level of proficiency relatively quickly implies a base level of training and professionalism, and a structure which enables the Army quickly to bring itself up to combat readiness. Both of these suggest that the Cold War experiences meant that the Army was relatively well trained and prepared for war. But the fact nevertheless remains that both 7 Armoured and 4 Armoured Brigades were not ready for war and required considerable individual and formation training in theatre before they were. Even then, certain aspects of the training programme appeared to be thin – for example, there were only two division-level exercises testing the ability of the force to fight as a coherent whole, one of which was a paper exercise, and the ability to exercise with a full logistics train was similarly limited.

Value and utility

The onset of the Gulf War reassured many over the continued value of both the armed forces and the Army. The end of the Cold War had raised doubts over the value of the armed forces in general – without the Soviet threat, what were the armed forces protecting the UK against? But it also raised more specific questions over the value of the Army relative to the other services. Indeed, initial thinking appeared to prioritise the Navy and Air Force over the Army – the former because it could be deployed flexibly overseas, the latter because the biggest military threat to the UK appeared to come from air attack. In contrast, the Army appeared to be difficult to deploy in any strength (not least because a strategic deployment had not been seriously envisaged since the 1960s), and the risk of invasion seemed so unlikely as not to be taken seriously.

The perceived utility of the armed forces had risen during the previous decade as a result in large part of the Falklands Conflict. The Gulf War did much not only to support that perception but to build on it. The picture given is one of startling military success. The House of Commons Defence Committee, for example, reported that 'Operation Granby was a signal victory and a remarkable achievement'.[28] This perception of military utility led not only to the Major government's belief that Britain's armed forces allowed it to 'punch above its weight' in world affairs, but created a permissive atmosphere in which deployments could take place. Thus the Major government also began to argue that permanent membership of the Security

Council created certain responsibilities for world peace and order, responsibilities which at times had to be met by the use of military power. Subsequent deployments in the former Yugoslavia may not have been solely a result of military successes in the Gulf, but the perceived utility of the British military and in particular the Army certainly facilitated such deployments.

Special forces

One of the more prominent features of the Army's performance in the Gulf was its use of special forces – prominent that is because of the attention it received and not necessarily because of its impact on military operations. The coalition commander, US General Norman Schwarzkopf, a Vietnam veteran, was publicly sceptical over the possible role of special forces. In contrast, the British military commander in the Gulf, General Sir Peter de la Billière, had been a member of and the commander of the SAS. De la Billière appears to have persuaded Schwarzkopf that special forces could prove valuable, not least in intelligence gathering.[29] They could also be used as spotters for laser guided bombs. Their most public role during the war, however, came in the so-called 'SCUD-hunt' in the Iraqi desert, an operation critical to the strategic success of the operation by helping to keep Israel out of the war. Subsequent to the conflict, the much publicised activities of an SAS troop Bravo Two Zero (featured in de la Billière's memoirs as well as numerous best-sellers) cemented the myth of the SAS and special forces. Whether their role was significant to the outcome of the military campaign is questionable, and the extent to which the success of the SCUD-hunt was down to them is again uncertain. But the important point is that the reputation of the SAS had been increased and its profile in the Army once again raised.

Contingency planning

The deployment of a large armoured force to the Gulf proved far from straightforward. The Ministry of Defence lacked a ready-made plan for such a deployment. Its contingency planning had concentrated on the deployment of smaller-scale forces, and much of the work to deploy the Brigade therefore had to be initiated from scratch. This task was further complicated by uncertainties over the length and final size of the deployment.[30] Further, it was clear that the Army did not possess a force geared to such a deployment and had therefore been forced to create one from an armoured brigade designed to fight in Germany as part of a larger British unit. 7 Armoured Brigade Group was an ad hoc creation which required additional support units to allow it to operate independently of a British divisional structure, additional combat units to bring it up to strength, and considerable equipment modifications to allow it to fight in the desert rather than in Europe. A number of units, including combat formations, were initially under-strength (sometimes very substantially so), and required the use of troops from other regiments to bring them up to combat strength. For example a troop of 17/21 Lancers had to be deployed with The Queen's Royal Irish Hussars, a company from the Grenadier Guards with the Royal Scots Dragoon Guards and four complete platoons from a variety of regiments with the Staffords.[31] As Bryan Watkins commented

> When the call came, not only was there no formed body of British troops in being which was more than even partly ready to respond to the demands of such an operation, but even those formations who were ultimately despatched and who were to fight so well had been cobbled together, with men and equipment being plundered from every corner of the British Army of the Rhine, leaving it emasculated. Had the Eastern threat to NATO not been in terminal decline, the consequences of leaving a gaping hole in the defences of the North German Plain scarcely bear thinking about.[32]

The move itself was not without its problems. Ships were loaded according to commercial rather than military priorities (i.e. reducing costs by ensuring maximum loads per ship and a minimum number of ships used). Loads for units were therefore split amongst a variety of ships, equipment arriving in batches rather than single loads. As a result considerable frustration and delay was experienced as units received only part of their equipment at a time. Delays in shipping complicated matters yet further, ships arriving in a different sequence from their departure, and as a consequence uncertainty developed over when ships would actually arrive. As a British staff officer later commented

> We were indeed fortunate not to have to fight immediately on arrival in the Gulf as many of the ships were loaded to suit the demands of the load masters and movement control staffs, with little thought given to the generation of effective combat power.

In sum, the contingency planning was inadequate to the task in hand: there were no plans and no dedicated capability for this scale of deployment to this theatre of war. In this respect Cold War planning, with its focus on BAOR within Europe and on low intensity conflict without, appears to have poorly prepared the Army for the Gulf War. But the ability to react and make do, to develop plans and create, deploy and fight a force speak much for the Army's flexibility and pragmatism. And it is here that the lessons of the Cold War, if not necessarily the previous two decades, came home. The Army's forte was never planning but in reacting and adapting; it was not so much in dedicated forces for dedicated missions, but in flexibility and pragmatism. The success of Operation Granby, coupled to uncertainties over what the post-Cold War world may still hold, seemed in the early 1990s to vindicate the Army's approach and demonstrate that the lessons of the Cold War were not only well learnt but appropriate. Of course, there are limits to what flexibility and an ability to adapt can accomplish, and there are arguments that in certain circumstances there is a benefit in having detailed plans and dedicated forces to hand. But with constrained resources meaning that the British Army would have to be more of a general force than one with dedicated units for specific tasks, it is not difficult to see why this approach caught the Army's favour.

Doctrine

British commanders in the Gulf were clearly very aware of doctrinal issues. Major-General Smith had been a director of the Higher Command and Staff Course, while a number of his senior commanders had been students on the course (including the commander of 7 Armoured Brigade, and the commander of the artillery group, Brigadiers Cordingley and Durie respectively). This gave commanders not only a shared intellectual base to work from, but a shared sense of the importance of doctrine and of thinking about how to fight the war. Doctrine and thinking about war were considered important by commanders in the Gulf, while after the war attention was paid to whether or not the doctrine had worked. What is also interesting is that where doctrinal inadequacies were identified, the response was to improve that doctrine rather than question the worth of doctrine.

The Gulf War also saw an attempt to fight a fully integrated, all-arms battle. When this was not achieved (for example with the integration of the deep artillery battle to the close battle) the fault was ascribed to technology rather than institutional reservations. Less successful was inter-service cooperation, especially attempts to integrate the land and air battles. Mitigating factors here included the use of American aircraft to support British land operations, and the lack of time in which to develop a joint doctrine. Nevertheless, Smith clearly attempted to fight a battle where all elements were integrated into a synergistic and coherent whole, and where all-arms cooperation functioned smoothly and efficiently.

Although the British were themselves unable to fight at what would normally be considered the operational level since only a single British division was deployed, they were able to fight as part of a force working at the operational level. A British staff officer working with CENTCOM for example commented that the operational vision of British commanders enabled them to cope with the scale of operations and influence planning much more than if they had operated at the tactical level. Further, Smith's divisional concept was a coherent plan whereby the two brigades and artillery group operated together rather than independently of each other, and the actions of each were seen as mutually supporting and as part of a grander design. In particular the manner in which the deep battle was to be integrated with the close suggests an awareness of the operational level of war. What is important is the emphasis placed by Smith upon fighting the *division* rather than simply a succession of firefights involving battle groups, and upon the development of a coherent concept of operations. All of this suggested a movement away from a tactical focus towards more of an operational-level focus. Similarly, the way in which Smith fought the division reflects a clear movement away from a preference for an attritional, set-piece approach to battle and towards one of manoeuvre. Manoeuvre was not simply the rapid movement of mass against an operationally static enemy, but the ability to act in a flexible manner. In other words it was as much an attitude of mind as a physical capacity. Therefore even though 1st Armoured Division advanced rapidly, taking the enemy by surprise and defeating them by manoeuvre as well as by weight of fire, what was also important was the command flexibility best seen in the exploitation beyond the initial target line (phase line SMASH at the Wadi al Batin) towards the main Kuwait City-Basra highway. Smith's emphasis upon seizing the initiative and maintaining a high tempo of operations was matched by his ability and willingness to act and react quickly, moving formations to exploit opportunities and changing his concept of operations when necessary.

The doctrinal reforms initiated by Bagnall and subsequently developed by the Army were seen to have been vindicated in the 1990–1 Gulf War. In particular the emphasis upon manoeuvre warfare was widely seen to have been a resounding success (although the fact that the desert was particularly suitable for fast mobile operations was also noted). Nevertheless, a number of deficiencies in doctrine were identified. First, manoeuvre warfare had increased the tempo of the battlefield, not only in terms of mobility but also in terms of decision making and reaction times, but the British Army had had little practical experience of handling large formations under such conditions. Much of its training in Germany and Canada had been at battle group level, not divisional. Second, although the British Army had attempted to move away from rigid and constricting orders towards a more flexible and mission-oriented style of command, these were not fully reflected in battlefield drills, which still tended to be rigid and formalistic. Third, fatigue proved to be a major problem, and more so than expected. The division was continuously engaged for sixty-six hours of the 100-hour war, with no respite for night or bad weather lest the momentum and tempo of operations be disrupted. By the end of the war troops were exhausted, and it appeared doubtful whether they could have continued high-tempo operations for much longer. It is tempting to suggest that the lack of a third brigade was at least partly responsible for this, keeping pressure on the front two brigades, but the manner in which Smith fought the division, using the brigades sequentially, appears to have dealt with some if not all of the problems created by the lack of a third brigade. Rather, fatigue was a product of the tempo of manoeuvre operations, of the need to maintain momentum and to stay inside the enemy's decision making cycle. Doctrine therefore needed to take account of these demands. Fourth, logistics doctrine had not been developed to match the requirements of manoeuvre warfare, and much had to be developed on an ad hoc basis in theatre. Fifth, joint doctrine with the air force was unsatisfactory, particularly as regards cooperation between artillery and close air support. Although a system of liaison officers was used, this could not compensate for the lack of joint doctrine. The impression is one of ad hoc measures rather than prearranged, well thought out and extensively practised drills. As Brigadier Durie, commander of the artillery group, commented

> Had the Iraqi resistance been stiffer, the lack of common doctrine, standard operating procedures and drills between the land and air forces could have been significant, and with more painful consequences.

This is ironic given the emphasis placed on land-air cooperation in formal statements on doctrine throughout the 1980s. It is evident that good intentions were not always realised in working practices. Although some of this may be explained by the fact that American not British aircraft provided much of the cover for British land forces, the failure to develop adequate procedures with the Americans suggests that the priority given to establishing adequate levels of cooperation in the Gulf did not match that given to it in the theory. Finally, the emphasis upon manoeuvre developed in peacetime had tended to obscure the importance of winning the firefight in war. Operation Granby brought home the importance of winning the firefight – or what was identified as the need for a 'violent storm'. As the Army's own account of operations comments, 'movement does not win battles unless it is used as a means of delivering firepower'. What these 'lessons' reveal, however, is that the British Army reinforced the theory of the 1980s with the practice of the Gulf War: manoeuvre warfare was seen to be vindicated, and the lessons learned concerned the better execution of such operations rather than fundamental problems suggesting a return to a more attritional style of warfare.

Conclusion

So how well did the later Cold War decades prepare the Army for the 1990–1 Gulf War? The answer, unsurprisingly perhaps, is mixed, but on the whole more positive than negative. What is also important to note is how well the Army *believed* the Cold War had prepared it both for the 1990–1 Gulf War and beyond. Here the answer is rather more positive. With the exception of certain discrete elements – such as the requirement to develop a strategic lift capacity – the Army did not see Granby as demonstrating the need for fundamental change. The emphasis upon human qualities whilst not ignoring machines, a key feature of the Army during the Cold War, was seen to have been vindicated. Praise for the quality and professionalism of its troops was widely echoed in the UK, whereas equipment failings were more heavily criticised. By inference, success was possible with deficient machinery, but contingent upon high-quality troops. Training regimes did not wholly prepare the Army for high intensity combat, but did provide a basis which could be built upon very effectively. The perceived utility of Britain's armed forces was reinforced, while their value, which had been questioned with the end of the Cold War, had been re-established. Special forces had once again received considerable attention and had performed valuable if not perhaps critical roles. Contingency planning had been poor, but the flexibility and adaptability of the Army – something Cold War insurgencies and before that imperial policing had impressed upon the Army – had been invaluable. And the new manoeuvrist doctrine had under-pinned the way in which 1st Armoured Division had fought so successfully. The success of 1st Armoured Division cannot of course be divorced from the wider coalition operation and in particular US ground and air power; but the fact that the small British division (just about) held its own with the best army in the world at this form of warfare, and was able to play an important role in a huge and hugely successful operation, suggests that at some very basic level the Cold War had prepared the Army well for Operation Granby.

Notes

1 This chapter draws on a range of sources in addition to those acknowledged in the notes. I am grateful to the Army's Tactical Doctrine Retrieval Centre for access to its excellent archive on the Gulf War and to a wide variety of individuals who agreed to be interviewed over the course of a number of years. None of these can be formally acknowledged in individual footnotes. I have also been fortunate enough to be a participant observer in a number of meetings, symposia, conferences and discussions with the British Army before, during and after the Gulf War. These have informed my views about developments over the past two decades but again cannot be acknowledged in individual footnotes.

2 'Small', that is, in terms of the percentage of the Army deployed and the nature of forces used – relatively light commando-type units. For the Royal Navy and for British defence policy of course, the Falklands Conflict was anything but small.

3 See for example Michael Dewar, *Brush Fire Wars: Minor Campaigns of the British Army since 1945*, London: Robert Hale, 1990; Thomas Mockaitis, *British Counterinsurgency in the Post-Imperial Era*, Manchester: Manchester University Press, 1995.

4 A defence review in all but name begun by the Conservative government in 1990 before the Gulf War, and with a White Paper on the future size and structure of the Army published after the Gulf War in 1991. For a discussion of the impact of *Options for Change* on the Army, see Mike Clark and Philip Sabin (eds) *British Defence Policy in the 1990s*, London: Brassey's, 1993, especially pp. 198–219.

5 *The Strategic Defence Review* Cm 3999, London: TSO, 1998. For a discussion of both the Strategic Defence Review and the 'rolling review' of the mid-1990s, see Colin McInnes, 'Labour's Strategic Defence Review', *International Affairs* vol. 74, no. 4 (1998), pp. 823–45.

6 See for example William Jackson and Edwin Bramall, *The Chiefs: The Story of the United Kingdom Chiefs of Staff*, London: Brassey's, 1992, pp. 351–421; Ritchie Ovendale, *British Defence Policy since 1945*, Manchester: Manchester University Press, 1994, pp. 131–57.

7 On the Dhofar campaign see Mockaitis, *British Counterinsurgency in the Post-Imperial Era*.

8 Northern Ireland primarily required infantry. Although other combat arms were deployed, these were mainly in the infantry role while the support role was very different from that of either BAOR or the Gulf War.

9 Although this was not sufficient to ward off a 'funding gap' in the late 1980s. On defence expenditure, see Third Report from the Defence Committee Session 1984–85, *Defence Commitments and Resources and the Defence Estimates 1985–6*, London: HMSO, 1985, especially vol. 1 pp. xvff.

10 An interesting reflection of this was the report of the House of Commons Defence Committee on the implications of the end of the Cold War. The Committee recognised that this was now a time of change, but was unable to think much beyond the reduced threat from the Soviet Union. Tenth Report from the Defence Committee Session 1989–90, *Defence Implications of Recent Events*, London: HMSO, 1990.

11 Tony Geraghty, *Who Dares Wins: The Story of the Special Air Service 1950–1980* (London: Arms and Armour, 1980). Tim Jones has done much to improve our understanding of the role of the SAS in immediate post-war counter-insurgencies. Tim Jones, *Post-War Counterinsurgency and the SAS, 1945–52: A Special Type of War*, London: Frank Cass, 2001.

12 The role of the SAS is detailed in Mark Urban, *Big Boys' Rules: The SAS and the Secret Struggle against the IRA*, London: Faber, 1993.

13 Hew Strachan, *The Politics of the British Army*, Oxford: Oxford University Press, 1997, pp. 188–9.

14 There is for instance no mention of special forces in General Sir Nigel Bagnall's seminal article introducing the new corps concept for BAOR. General Sir Nigel Bagnall, 'Concepts of Land/air Operations in the Central Region I', *Journal of the RUSI*, vol. 129, no. 3 (1984), p. 60.

15 B. Holden Reid, 'Introduction: Is there a British Military "Philosophy"?', in Major-General J.J.G. Mackenzie and B. Holden Reid (eds) *Central Region Vs. Out of Area: Future Commitments*, London: Tri-Service, 1990, p. 1.

16 For a discussion of this, see Colin McInnes and John Stone, 'The British Army and Military Doctrine', in Michael Duffy, Theo Farrell and Geoffrey Sloan (eds) *Doctrine and Military Effectiveness*, Strategic Policy Studies 1, Exeter: Strategic Policy Studies Group, University of Exeter, 1997, pp. 14–25.

17 Quoted in Williamson Murray, 'British Military Effectiveness in the Second World War', in Allan R. Millett and Williamson Murray (eds) *Military Effectiveness, vol. III: The Second World War*, London: Allen and Unwin, 1988, p. 111.

18 *Design for Military Operations: British Military Doctrine*, Army Code 71451 (1989), usually referred to as *British Military Doctrine* or *BMD*. Although subtitled *British* Military Doctrine, this was a single-service document by and for the Army.

19 Major-General J.J.G. Mackenzie and Brian Holden Reid (eds), *The British Army and the Operational Level of War*, London: Tri-Service, 1989, p. ix.

20 Bagnall, 'Concepts of Land/air Operations'; Colin McInnes, *Hot War, Cold War: The British Army's Way in Warfare 1945–95*, London: Brassey's, 1996.

21 For an official and reasonably detailed account of 1st Armoured Division's preparation for and conduct of operations, see *Operation Desert Sabre: The Liberation of Kuwait 1990–1*, Army Code 71520 (1993). See also Nigel Pearce,

The Shield and the Sabre: The Desert Rats in the Gulf 1990–1, London: HMSO, 1992.

22 *Statement on the Defence Estimates 1992* CM1981, London: HMSO, 1992, p. 78.
23 Tenth Report from the Defence Committee Session 1990–1, *Preliminary Lessons from Operation Granby*, London: HMSO, 1991, p. vii.
24 Ibid.
25 Major General Patrick Cordingley, *In the Eye of the Storm*, London: Hodder and Stoughton, 1996, p. 18.
26 *Preliminary Lessons from Operation Granby*, p. xiv.
27 Cordingley, pp. 62ff.
28 Fifth Report from the Defence Committee Session 1993–4, *Implementation of Lessons Learned from Operation Granby*, London: HMSO, 1994. See also Dispatch by Joint Commander of Operation Granby, *London Gazette* 29 June 1991.
29 General Sir Peter de la Billière, *Storm Command*, London: HarperCollins, 1992.
30 *Preliminary Lessons from Operation Granby*, p. xiii.
31 See for example Nicholas Benson's account of the Staffordshire Regiment in the Gulf. Nicholas Benson, *Rats' Tales: The Staffordshire Regiment at War in the Gulf*, London: Brassey's, 1994.
32 Bryan Watkins, 'The Aftermath of the Storm?', *Journal of the RUSI*, vol. 138, no. 3 (1993), p. 52.

Index

Printed in Great Britain
by Amazon